01-CTS-32ET

A NEW TOOL FOR MEASURING BIOSOLIDS FLOC STRENGTH

by:

Mohammad M. Abu-Orf
Veolia Water North America

Banu Örmeci
Department of Civil and Environmental Engineering
Duke University

2004

Water Environment Research Foundation

The Water Environment Research Foundation, a not-for-profit organization, funds and manages water quality research for its subscribers through a diverse public-private partnership between municipal utilities, corporations, academia, industry, and the federal government. WERF subscribers include municipal and regional water and wastewater utilities, industrial corporations, environmental engineering firms, and others that share a commitment to cost-effective water quality solutions. WERF is dedicated to advancing science and technology addressing water quality issues as they impact water resources, the atmosphere, the lands, and quality of life.

For more information, contact:
Water Environment Research Foundation
635 Slaters Lane, Suite 300
Alexandria, VA 22314-1177
Tel: (703) 684-2470
Fax: (703) 299-0742
www.werf.org
werf@werf.org

This report was co-published by the following organizations. For non-subscriber sales information, contact:

Water Environment Federation
601 Wythe Street
Alexandria, VA 22314-1994
Tel: (800) 666-0206
Tel: (703) 684-2452
Fax: (703) 684-2492
www.wef.org
pubs@wef.org

IWA Publishing
Alliance House, 12 Caxton Street
London SW1H 0QS, United Kingdom
Tel: +44 (0) 20 7654 5500
Fax: +44 (0) 20 7654 5555
www.iwapublishing.com
publications@iwap.co.uk

Library of Congress Catalog Card Number: 2003115649

IWAP ISBN: 1-84339-713-7

Veolia Water North America, Duke University

ABSTRACT AND BENEFITS

Abstract:

The ability to measure sludge network strength is important in sludge dewatering applications because it can be used to determine the optimum polymer dose for conditioning to achieve good dewaterability. This was demonstrated through laboratory and full-scale dewatering and thickening. The network strength increased as the polymer dose was increased. However, at the optimum dose a "drop" in the network strength occurred. Further research is needed to verify this concept at full scale and to provide a robust technology to the water and wastewater treatment industry. A second phase will be sought through WERF funding.

Rheometry was used to determine the sludge network strength. This report also formulated and demonstrated a standard protocol for measuring network strength in terms of energy dissipated in a certain volume of sludge. This report describes two protocols for measuring network strengths by either torque or concentric cylinder rheometers. A mathematical derivation has shown that area under rheograms, namely the curves which were developed by plotting shear rate (1/sec) versus shear stress (Pa) and time (sec) versus torque (mNm), indicated the rate of energy dissipation within the sludge system and the total dissipated energy was related to the network strength.

The research did not intend to measure the "absolute" network strength, rather a comparative strength of different aggregates using the same instrument and under the same measuring conditions. Thus, the measured strength is relative.

Benefits:

- Presents a new method for determining the optimum polymer dose for sludge conditioning prior to dewatering and thickening. A drop in the network strength indicates the optimum polymer dose. This has important implications for water and wastewater treatment plants seeking to reduce polymer consumption and automate dewatering.
- Presents two methods and mathematical models for estimating the sludge network strength using two rheometry concepts.
- Verifies the reproducibility of the rheological measurements using the torque rheometers.
- Correlates the network strength to the polymer dose used to condition sludge; the network strength increased as the polymer dose was increased up to the level at which a drop in the network strength was observed.
- Presents a new method for determining the optimum polymer dose for conditioning sludge prior to dewatering and thickening. A drop in the network strength indicates the optimum polymer dose. This has important implications for water and wastewater treatment plants seeking to reduce polymer consumption and automate dewatering.

- Presents an evaluation of the factors that affect sludge network strength, including polymer type, mixing conditions, rotational speed within the rheometer, and inorganic conditioners.

Keywords: Polymer monitoring, network strength, conditioning, dewatering, biosolids, sludge, rheology

ACKNOWLEDGMENTS

The authors wish to thank the Water Environment Research Foundation (WERF) and its subscribers for supporting this research, and hope that this document will provide an appropriate return on the investment. WERF's Project Advisory Committee provided invaluable assistance, as did Roy Ramani, Ph.D., the project manager. The authors of this report received a significant amount of help in completing this ambitious project. Firstly, U.S. Koei Industries personnel (notably Taka Okutsu/U.S. Koei) loaned us the instruments necessary to conduct this research, and greatly facilitated the on-time start up of the bench-scale tests. Secondly, the personnel at the Wilmington Pollution Control Facility provided valuable assistance in our full-scale test. Steve Dentel of the University of Delaware provided the Haake rheometer and assisted in many technical aspects of understanding rheology. Finally, the generous matching support by Veolia Water North America and facilities furnished by Duke University allowed this work to be well-performed.

Report Preparation

Principal Investigator:

Mohammad M. Abu-Orf, Ph.D.
Veolia Water North America

Co-Principal Investigator:

Banu Örmeci, Ph.D.
Department of Civil and Environmental Engineering, Duke University

Project Team:

Kidon Cho
Duke University

Project Subcommitteee

John Novak, Ph.D.
Virginal Polytechnic Institute and State University
Tata Prakasam, Ph.D
Tata Associate International, Inc.

Water Environment Research Foundation Staff

Director of Research: Daniel M. Woltering, Ph.D.
Project Manager: Roy Ramani, Ph.D.

TABLE OF CONTENTS

LIST OF TABLES

LIST OF FIGURES

LIST OF ACRONYMS

Note: Abbreviations for common English or SI units (mL, ppm, kV, etc.) have not been included.

CST	capillary suction time
DS	dry solids
GBT	gravity belt thickener
gpm	gallons per minute
M/D	monovalent to divalent cation ratio
MGD	million gallons per day
NATC	North American Technology Center
network	raw or conditioned sludge suspension
rpm	number of revolutions per minute
SD	standard deviation
TS	total solids
TTQ	totalized torque
TTF	time to filter
WAS	waste activated sludge
WERF	Water Environment Research Foundation
WWTP	wastewater treatment plant

EXECUTIVE SUMMARY

This project used the principles of rheology to formulate a protocol for measuring the floc strength of different types of polymer-conditioned sludge solids. The ability to measure floc strength is important in the dewaterability of many sludge applications. In this report we refer to raw or conditioned suspensions as "network." Thus, measurements of floc strength indicate the network strength, which is most important in sludge dewaterability applications. The research also used the network strength to obtain the optimum polymer-conditioning dose to achieve good dewaterability of sludges. The appropriateness of using network strength to determine optimum polymer dose for dewatering applications was demonstrated in laboratory and full-scale dewatering and thickening applications. However, more research should be conducted to test the technology in full-scale and control the polymer dose at the optimum level.

This report provides two protocols for measuring network strengths using either torque or concentric cylinder rheometers. Two mathematical models were used to calculate network strength. The models showed that the area under the rheograms (shear rate vs. shear stress or torque time relationship) indicated the rate of energy dissipation within the conditioned sludge solids matrix and the total dissipated energy were related to network strength. Measuring the sludge network strength in terms of energy is appropriate since the ability of the network to withstand shear forces depends upon the chemical and physical bonds that hold it together. The research did not intend to measure the "absolute" network strength, but measured the comparative network strength of different aggregates at different polymer doses using the same instrument, under the same measurement conditions. Thus, the measured network strength is relative.

Laboratory tests were designed to:

- Establish the reproducibility of the rheological measurements.
- Determine the effect of impeller speed on network strength.
- Identify the correlation between the polymer dose and network strength.
- Study the effect of mixing conditions during sludge conditioning on the network strength.
- Analyze the effect of specifically selected polymers on network strength.

Laboratory tests used four sludge types: anaerobically digested, aerobically digested, activated, and alum using both a torque and concentric rheometer. Good reproducibility of the rheology measurements was achieved using the torque rheometer. The reproducibility of results obtained with the concentric rheometer was inferior, and its use is not recommended for sludge applications—specifically when conducting measurements on conditioned, flocculated suspensions.

Laboratory experiments incorporated jar testing procedures. Capillary suction time and filtration tests were used to measure dewaterability, and the whole conditioned sample was used in rheology measurements. The calculated network strength increased as the polymer dose was increased. However, at the optimum dose a "drop" in the network strength occurred. The optimum impeller speed in the torque rheometer is the minimum that can produce a significant signal, however, this optimum speed depends on the solids content of sludge and increases with increased solids content. The network strength of specific sludges can be altered through changing the mixing conditions and polymer. This aspect is important in full-scale application as operators might need to change the network strength to achieve good dewatering with a particular piece of equipment such as a belt filter or centrifuge.

Full-scale experiments were conducted at the Wilmington Pollution Control Facility in Wilmington, Delaware. The tests were conducted on anaerobically digested sludge during centrifugation, and on mixed primary and waste activated sludge during gravity belt thickening. The dose-response procedures were used in the full-scale tests. All tests were conducted on-site to eliminate the possibility of alterations in the sample characteristics. Capillary suction time and solids analysis (filtrate/centrate and cake solids) were used to test the performance on conditioning and dewatering. Similar to the laboratory tests, the calculated network strength from the measurements increased with polymer dose, and at the optimum dose a "drop" in the network strength occurred.

CHAPTER 1.0

BACKGROUND AND PREVIOUS STUDIES

1.1 Overview

The ability to measure floc strength is important in sludge dewatering applications. Previous research (Tambo and Hozumi, 1979; Higgins and Novak 1997a,b) has focused on measuring sludge floc strength indirectly through measuring other parameters, such as floc size, floc density, and sludge dewaterability. Direct measurements of floc strength reported in the literature are not applicable to sludge, especially if it is conditioned, because these measurements are to be made on the flocs at the microscopic level. Throughout this report we will refer to raw or conditioned sludge suspension as "network," and thus measurements will indicate the network strength, which is the most important measurement in sludge dewaterability applications.

The research uses rheology, which is a fundamental character of sludge that can be used to develop a standard protocol for measuring network strength in terms of energy required to break up the structure of a certain volume of sludge. This report provided two protocols for measuring network strength: torque and concentric cylinder rheometers. These methods are the first attempt to *directly* measure sludge network strength with the use of such apparatus in terms of energy since the ability of the network to withstand shear forces depends upon the chemical and physical bonds that hold it together. The research did not intend to measure the "absolute" network strength. Rather, the intent was to measure the relative network strength of different floc aggregates obtained at different polymer dosages using the same instrument and under the same measuring conditions. Thus, the measured network strength is relative.

1.2 Terminology

The term "floc" has different annotations. Some referred to floc as the single unit in the suspension, while others referred to it as the whole conditioned suspension. To be consistent in terminology, we will refer to the raw or conditioned sludge suspension as "network" and thus measurements of floc strength will indicate the network strength. We followed the model of flocculation of kaolin suspensions offered by Michaels and Bolger (1962) in which they refer to the floc and not the single particle as the basic unit of flow under shear. The flocs tend to cluster together forming weak aggregates, which in turn grow by collision and break down by shear and thus forming **extended networks** that give the suspension its structural properties. Yen et al. (2002) also used the term network to describe flocculated material from activated sludge systems.

Again, we did not measure the "absolute" network strength, but only the relative network strengths of different aggregates using the same instrument and measurement conditions. Thus, the measured network strengths are relative.

1.3 Statement of Importance

Measuring sludge floc strength is an important aspect of sludge characterization. Current "floc" strength measurements are only empirical and based solely upon the final size of flocs once they break up with high mixing energy. Other floc strength measurements require complicated devices that are not practical for wastewater plant operations. The literature reveals two available categories for measuring floc strength.

The first category involves using other sludge properties to indirectly determine floc strength. One of the earliest attempts to measure floc strength came from Hannah et al. (1967) who measured the change in particle size distribution to estimate floc strength when applying a known amount of shear. Higgins and Novak (1997a,b) used the capillary suction time (CST) test results to characterize the ability of flocs to resist deterioration from shear forces. Other researchers (Tambo and Hozumi, 1979) used the floc size and density as an indication of floc strength.

The second category directly measures floc strength. Glasgow and Hsu (1982) developed a model and apparatus for measuring individual floc strength from the aggregation of colloidal kaolin and either ferric chloride ($FeCl_3$) or polymer. Yeung and Pelton (1996) developed micromechanical techniques that pull apart single floc particles (6-40 μm) and determine floc strengths. Although these techniques and methods are of great scientific value, they are not applicable to sludge operations. Measuring individual floc strength microscopically is not practical for sludge dewatering operations and cannot be used on a continuous basis.

Rheological properties can be used to understand the structure of flocculated suspensions due to their high sensitivity (Otsubo, 1996). These properties describe the deformation (strain) of a body under the influence of stresses. This report addresses the use of rheology for determining a suspension's strength, which is the first attempt to *directly* measure sludge network strength since the ability of a network to withstand shear forces depends upon the chemical and physical bonds that hold it together. One important aspect of measuring network strength is that effective sludge dewatering depends on the strength of the aggregates formed following polymer conditioning. Different network strengths are required for optimum dewatering when different dewatering devices are used. Novak and co-workers (Novak and Haugan, 1979; Novak et al., 1988; Novak, 1991; Novak and Bandak, 1989, 1994) pioneered the relationship between mixing parameters (velocity gradient [G] and time [t]) and polymer dose and indicated that to properly condition sludge for mechanical dewatering, the type of sludge and shear in the dewatering device must be considered. More recent research from Novak and colleagues suggests that monovalent to divalent (M/D) cations ratio in sludge significantly impacts floc strength and dewaterability (e.g., Novak et al., 1998; Higgins and Novak, 1997a,b).

The highest attainable network strength is not necessarily desirable for particle separation by sedimentation or filtration (Hannah et al., 1967). For example, centrifuges require higher network strength than belt filter presses, and in certain cases different levels of network strength are required for the same type of dewatering device. But, stronger networks would be more difficult to dewater (Glasgow and Hsu, 1982). *Thus, the required network strength value for producing a good dewaterable sludge with various dewatering devices is not well defined, and cannot be measured precisely with the current available methods and technologies.*

Moreover, there is *no direct* method to continuously measure the required strength of the flocculated material prior to its entering the dewatering device to assure the appropriate network strength that *matches* the shear within a given dewatering device or solids separation operation. In certain industrial solid–liquid separation processes, the network strength may be the main parameter for achieving good separation via dewatering devices.

Yen et al. (2002) used the power input to the system to derive and calculate the total and specific energy input to the suspension, where the specific energy input was defined as the energy to break down the flocculated network containing unit weight of solids. They indicated that the high specific energy input corresponds to a strong flocculated matrix. However, these researchers used the hysteresis loop principle in rheology to determine this specific energy by subtracting the energy input during the shear rate-decreasing phase from shear rate-increasing phase. Yen et al. (2002) used a concentric cylinder rheometer and offered no data concerning the reproducibility of the rheograms, even though it has been documented (Abu-Orf and Dentel, 1999) that the reproducibility of measurements with concentric rheometers is a major concern even when using synthetic sludge. Yen et al (2002) indicated that a sufficient space (gap between inner and outer cylinder) is essential in forming a comprehensive network, while rheological tests with narrowly spaced channels yield erroneous polymer dose assessments. The gap in concentric cylinder rheometers is usually narrow. Yen et al. (2002) found a correlation between network strength and polymer dose. In almost all cases presented, increasing the polymer dose beyond the under-dosing region (range below optimum polymer dose) resulted in increase in the specific energy or network strength. However, the data strongly suggest that the specific energy cannot be used to determine the optimum polymer dose since an almost linear relationship exists between specific energy and polymer dosage with no observed change in the specific energy or network at the optimum dose.

1.4 Research Objectives

The objective of this research was to use certain rheological measurements to arrive at a standard protocol to *directly* measure the sludge network strength in terms of energy dissipated in a given volume of a suspension. Another objective was to correlate the network strength to sludge dewaterability. We hypothesized that the suspension's strength would increased with a higher polymer dose. However, at the optimum polymer dose this strength would be lower, which would allow the network to release some of the incorporated water while maintaining its structural integrity. This would lead to good suspension dewaterability at the optimum dose without complete network deterioration.

1.5 Methods for Estimating Relative Sludge Network Strength Using Rheological Measurements

1.5.1 Use of Concentric Rheometers

These rheometers measure the rheological properties of non-Newtonian suspensions. The measurements are conducted by increasing the shear rate from zero to a predetermined value over a given time interval. The measurement can be conducted on flocculated and non-flocculated suspensions. The instrument measures the shear stress of the network at the different shear rates as shown in Figure 1-1a, which represents typical output of a concentric rheometer. The measured/reported values incorporate the correction factors for the sensors used, so the results should be independent of the sensor's type.

We used the area under the rheogram (shear stress vs. shear rate graph, Figure 1-1a) to estimate the network relative strength. This concept is clarified below.

For a Newtonian fluid that is sandwiched between two parallel plates, Newton's law of viscosity states that:

$$\tau = \frac{F}{A} = \mu \frac{dv}{dy} \qquad (1.5.1\text{-}1)$$

where τ is the shear stress, F is the force parallel to the plates, A is the area of each plate, μ is the viscosity of the fluid, and dv/dy is the velocity gradient or shear rate (γ) of the fluid. If we multiply both sides of the equation by dv/dy, we arrive at the following equation:

$$\left(\frac{F}{A}\right)\left(\frac{dv}{dy}\right) = \mu\left(\frac{dv}{dy}\right)^2 \qquad (1.5.1\text{-}2)$$

If we write the velocity v as dx/dt:

$$\frac{d\left(\frac{Fdx}{Ady}\right)}{dt} = \mu\left(\frac{dv}{dy}\right)^2 \qquad (1.5.1\text{-}3)$$

Since Fdx equals an increment of energy dE, and Ady equals an increment of volume dV, equation 1.5.1-3 becomes:

$$\frac{(dE/dV)}{dt} = \mu\left(\frac{dv}{dy}\right)^2 \qquad (1.5.1\text{-}4)$$

The quantity dE/dV is the *energy dissipation per unit volume and* (dE/dV)/dt represents *the rate of energy dissipation per unit volume* (Hiemenz and Rajagopalan, 1997).

For Newtonian fluids, the shear stress is proportional to shear rate, and the viscosity μ of the fluid is the factor of proportionality. The area under this rheogram can be calculated as follows:

$$Area = \int\left(\frac{F}{A}\right)d\left(\frac{dv}{dy}\right) = \int\mu\left(\frac{dv}{dy}\right)d\left(\frac{dv}{dy}\right) = \frac{\mu}{2}\left(\frac{dv}{dy}\right)^2 \qquad (1.5.1\text{-}5)$$

Thus comparing equations 1.5.1-4 and 1.5.1-5,

$$Area = \frac{1}{2}\left(\frac{dE}{dV}\right)dt \qquad (1.5.1\text{-}6)$$

This shows that the area under a shear stress-shear rate rheogram of a Newtonian fluid is linearly proportional to the rate of energy dissipation per unit volume. *Thus, the area under the shear stress-shear rate graph represents the rate of energy dissipation per unit volume.*

How is the energy dissipation in the flocculated sludge network related to the network strength? Michaels and Bolger (1962) noted three forms of energy dissipation during flocculation when shear forces were applied: the network energy, the aggregate or creep energy, and the viscous energy.

Under shear forces, the network bonds within a suspension are stretched and deformed. When shear exceeds the yield stress of any network link, the link will break and the broken link will snap back to its unstrained position. This leads to local viscous oscillation, where energy is dissipated in a fashion similar to a spring, which has been stretched by a cord that is broken. This energy was termed network energy. The aggregate or creep energy is dissipated under further shear where more energy is used to break down aggregate structure by breaking the bonds between the flocs. Finally, viscous energy is dissipated by fluid flowing around the flocs where viscous energy losses occur due to the presence of particles. *Thus, the total energy dissipation into the suspension represents the network strength.*

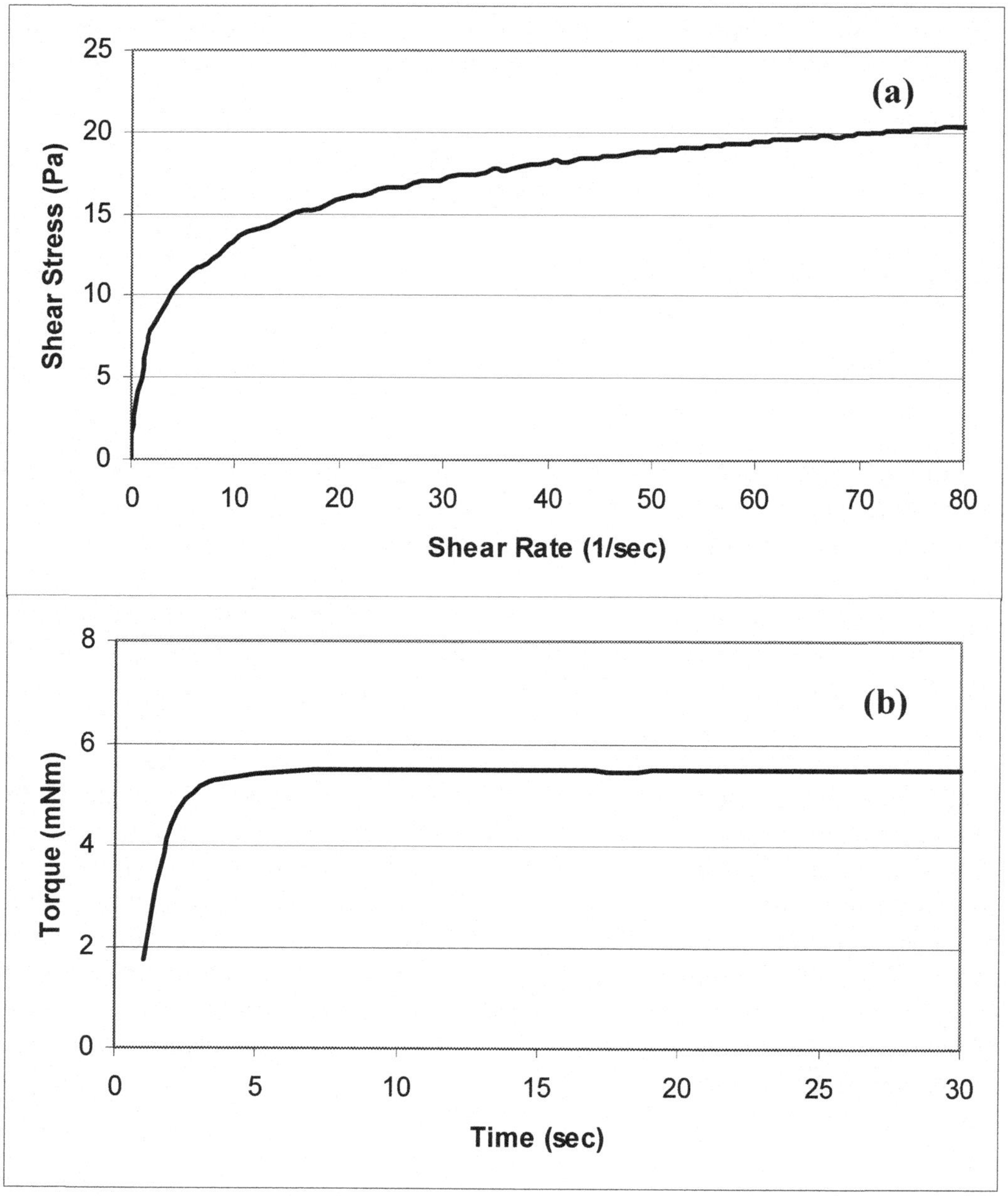

Figure 1-1. Typical Shear Stress-shear Rate and Torque-time Rheograms for Sludge.

Knowing the time required to reach a specific shear rate, the rate of energy dissipation per unit volume can be converted to total energy dissipation per unit volume in the network. This is accomplished by multiplying the area under the shear stress vs. shear rate rheogram with the required time to reach that specific shear rate. This equals the total energy dissipation per unit volume in the network, and represents the network strength. The following example further explains this relationship.

Assume that the area under the rheogram up to a shear rate of 25 sec^{-1} equals A. If the time to reach shear rate of 25 sec^{-1} is t (sec), then the total energy dissipation (E_{td}) in the network per unit volume equals $E_{td} = 2A\,t$. If the solids content of the sample is X (kg/m^3), then:

$$E_{td} = \frac{2A\,t}{X} \qquad (1.5.1\text{-}7)$$

Where,

E_{td} = total energy dissipation per unit of dry solids (DS) (Joule/ kg DS) and represents network strength;

A = area under the rheogram up to a certain shear rate ($kg/m.sec^3$ or Pa/sec);

t = time to reach a specific shear rate (sec); and

X = solids concentration of the sample (kg/m^3).

For non-Newtonian fluids that follow the Bingham plastic model:

$$\frac{F}{A} = \mu_p\left(\frac{dv}{dy}\right) + \tau_o \qquad (1.5.1\text{-}8)$$

where μ_p is the plastic viscosity, and τ_o is the yield stress.

Similar to the derivation of equation 1.5.1-4, the rate of energy dissipation per unit volume can be obtained by multiplying both sides of the equation by dv/dy:

$$\left(\frac{F}{A}\right)\left(\frac{dv}{dy}\right) = \left(\frac{dE}{dV}\right)/\,dt = \mu_p\left(\frac{dv}{dy}\right)^2 + \tau_o\left(\frac{dv}{dy}\right) \qquad (1.5.1\text{-}9)$$

The area under the shear stress-shear rheogram can be calculated as follows:

$$Area = \int\left[\mu_p\left(\frac{dv}{dy}\right) + \tau_o\right] d\left(\frac{dv}{dy}\right) = \frac{1}{2}\mu_p\left(\frac{dv}{dy}\right)^2 + \tau_o\left(\frac{dv}{dy}\right) \qquad (1.5.1\text{-}10)$$

Comparison of equations 1.5.1-9 and 1.5.1-10 shows that the area under a shear stress-shear rate rheogram of a non-Newtonian Bingham plastic fluid represents the rate of energy dissipation per unit volume. It is important to point out that the rheogram area will not be linearly proportional to the rate of energy dissipation as is the case for Newtonian fluids.

When comparing two network structures, at a given shear rate on the rheogram, if the area under the rheogram of network A is greater than the area of network B, then the energy dissipated in network A is greater than the energy dissipated in network B. This means that A has a stronger network structure than B. *Thus, the area under the shear stress vs. shear rate rheogram is a measure of relative network strength.*

1.5.2 Use of Torque Rheometers

Torque rheometers were developed during the 1930s and were used to characterize the uniformity of flour dough mixing by the baking industry. Currently, they are widely used to study consistency, processability, and rheological properties in polymer melts in the plastic and rubber industries. Torque rheometers characterize the flow behavior of a suspension by measuring the viscosity-related torque as a result of the suspension's resistance to shear. The evolution of torque rheometers moved from mechanical to the electronic instruments currently in use.

Torque rheometers most commonly apply a constant shear in terms of rotational speed (N, rpm) to the sample. They measure the resistance of a sample to the shear applied as strain by a transducer. This strain is translated to an electronic signal, which is converted to units of torque, M, (either, g-m, or N-m). The detected torque values are reported as a function of the time the shear is applied. The plot of the test results is called a torque rheogram (time vs. torque graph). Since torque rheometers utilize larger sample volumes and various shapes and complex geometry of mixing rotors, flow patterns and correlation factors are different than those of the traditional concentric rotational rheometers. As a result, direct conversion of torque values to shear stress and shear rate may not be very accurate and is not suggested by the manufacturers.

In general, (Cheremisinoff, 1986):
γ (mean shear rate) $= C_1 N$
τ (mean shear stress) $= C_2 M$
Where,
C_1 and C_2 are constants depending on the geometry of the mixer rotor and cup;
N = number of revolutions per minute (rpm); and
M = torque.

Thus, the measured torque multiplied by the rotational speed, which is the area under the torque-time rheogram, is also a representative of the total energy dissipation rate in a suspension. Figure 1-1b shows a typical rheogram of torque-time relationship. Cheremisinoff (1986) defines the area under the torque-time curve at a given time as "totalized torque" (TTQ), which represents the energy that goes into the system for that given time. TTQ values provide information for comparison purposes between different samples. For example, torque rheograms of two samples may exhibit very different shapes but stabilize at the same torque value at the end of the test. If we compare only the final torque values, the samples may seem identical. However, comparison of TTQ values reveals the differences in the network strength of the samples.

The area under the torque-time rheogram can be used to estimate the network relative strength after converting the TTQ to work energy as described in the following discussion (modified from Cheremisinoff, 1986).

Let's assume that the impeller has a radius, r. The impeller travels a distance equal $2\pi rN$, where N is the revolutions per minute. The work energy (W) then equals force (F) multiplied by the distance covered by the impeller. Thus, $W = 2\pi rNF$. Since torque (M) = F r

The total work energy (W_t) = $2\pi r N \int_{t1}^{t2} (M(t)/r)\,dt$

$W_t = 2\pi N \times TTQ$

And, the unit work energy is

$$W_u = 2\pi N \times (TTQ)/V_b \qquad 1.5.2\text{-}1$$

Where,

W_u = unit work energy and has units of (Joule/cm^3);

N = rotational speed in rpm;

TTQ = totalized torque, area under the rheogram and has units of (Joule-min); and

V_b = the processed volume samples in cm^3.

The unit work (W_u) is defined as the work energy required for processing a unit volume. As previously discussed, and according to Michaels and Bolger (1962), the energy required to process a given mass or volume of a material is representative of the energy dissipated into the suspension which indicates the suspension's network strength.

Using either rheometer method, it is important to point out that the proposed methods are not measuring the "absolute" network strength, but a comparative strength of different aggregates using the same instrument and measurement conditions. Thus, the measured strength is relative. Moreover, our reported network strength values are cumulative of the network strength over time.

The derivations illustrated above show that the area under a torque-time or a shear stress-shear rate rheogram is representative of the energy dissipation into the suspension and can be used to calculate the suspension's network strength. This is a very useful tool for comparison of multiple samples with different characteristics. Determination of network strength from rheological measurements can be used to identify the differences between samples or to assure all samples have identical properties. In some cases, determination of network strength may not even be needed since comparison of samples can simply be achieved by directly comparing the areas under the rheograms. For comparison of flocculated suspensions, it is suggested to use the torque rheometers instead of the concentric-cylinder rheometers. Torque rheometers are rapid in use, simple to operate and generate accurate, reproducible results. In addition, they employ larger sample volumes compared to concentric-cylinder rheometers, which helps minimize the problems experienced with the rheological measurements of flocculated suspensions. Problems associated with taking representative sub-samples from well-flocculated sludges with two distinct phases of solids and water were reported previously (Abu-Orf and Dentel, 1999). Torque rheometers also offer the advantage of adding and mixing conditioners directly in the sample, and using the whole sample volume for the rheological measurements. In addition, the calculation of network strength from the torque-time rheogram is very simple and straightforward as illustrated in equation 1.5.2-1, and applies to both Newtonian and non-Newtonian fluids.

CHAPTER 2.0

MATERIALS AND METHODS

This research focuses on using torque rheometers to estimate sludge network strength. *However, use of concentric cylinder rheometers is also assessed; these methods and results are presented in Appendix A.*

2.3 Torque Rheometer Components

Flockky Tester (Figure 2-1), a torque rheometer manufactured by Koei Industry Co., Ltd. (Japan), was used in these experiments. The Floccky Tester has two main components: the detection unit and the control unit. The detection unit consists of a motor, impeller, and a load-cell, and is connected to the control unit. The control unit is connected to and operated by a personal computer. All parameters are selected through the system's software, including the rotational speed of the detector, duration of the rotation, and the sensitivity of the detector.

The detection unit includes a cylindrical jar where 200 mL of sludge is sheared with a three-vane impeller (Figure 2-2). The load cell detects the resistance of the suspension to the shear created by the rotating impeller. Thus, the rheometer measures the changes in the viscosity of the slurry sample that is stirred with a rotating detector. The resistance is measured in terms of electrical signals, and the data are reported in voltage (mv) or torque (mNm). The impeller speed can be varied from 30-420 revolutions per minute (rpm). The beaker has two side ports, through which conditioners can be delivered directly to the suspension in the beaker, which allows a consistent conditioner delivery (Figure 2-3).

Floccky software reports the following information supplementary to the rheogram.

- Peak (mv or mNm): Indicates the maximum signal level between vertical slice lines.

- Sum: Indicates the value of the area under the rheogram that is between horizontal and vertical slice lines. There are four slice lines, two vertical and two horizontal. The slice lines determine the boundaries of the data to be included in the analysis.

2.2 Sludge Samples

Four different types of sludges were used in the experiments: alum sludge, activated sludge, aerobically digested sludge, and anaerobically digested sludge. Activated sludge and anaerobically digested sludge samples were collected from North Durham Water Reclamation Facility (Durham, NC). Alum sludge samples were collected from Williams Water Treatment Plant (Durham, NC). Aerobically digested sludge samples were collected from South Durham Water Reclamation Facility (Durham, NC).

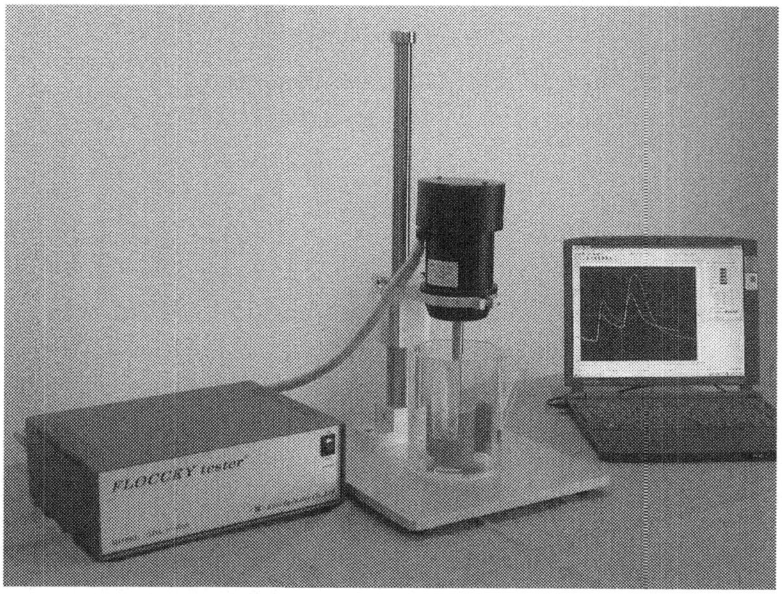

Figure 2-1. Main Components of the Floccky Rheometer.

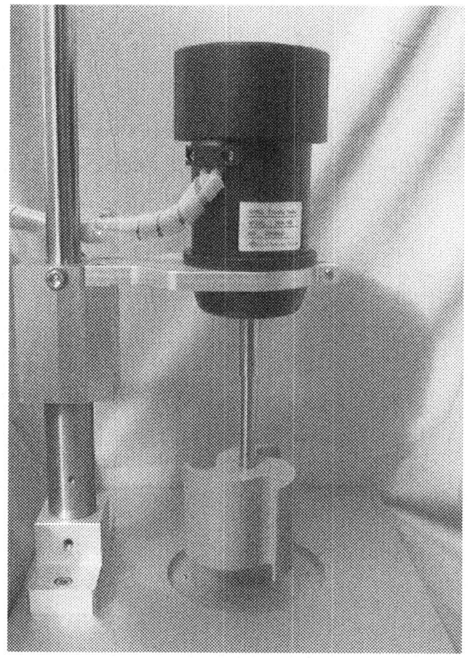

Figure 2-2. Rheometer Detector and Impeller.

Figure 2-3. The Beaker with Two Ports for Delivering Conditioners.

2.3 Polymer Conditioning—Jar Test Methods

Sludge samples of 200 mL were conditioned with various polymers using a jar tester (Stockhousen 655K, Superfloc SD-2080, Superfloc SD-2081, Superfloc SD-2083, Superfloc SD-2085, C-1598). These samples were directly used in rheological and dewaterability measurements without the use of subsamples. This eliminated the variation in measurements caused by the differences between subsamples as observed by Abu-Orf and Dentel (1999). Different combinations of mixing conditions, such as rotational speeds and mixing times, were used for testing.

2.4 Dewaterability Measurements

CST tests were conducted in triplicate according to the *Standard Methods* (APHA, 1995). For the filtration tests, the 200 mL conditioned sludge sample was vacuum-filtered for 10 minutes. The filtrate volume was measured and the cake solids collected on the filter paper were analyzed for total solids content according to the *Standard Methods*. This modified filtration testing eliminated differences in sub-sampling experienced during time to filter (TTF) test and allowed measurement of the unfiltered cake solids. Triplicates were used for both the CST and the filtration measurements.

2.5 Rheological Measurements

The torque rheometer uses a standard 200 mL sample volume that is placed in the unit's beaker prior to measurement. All rheological measurements were conducted in triplicate using three different sludge samples from the same batch for 2 minutes (unless otherwise stated). After each measurement, the sample was discarded due to the thixotropic nature of sludge. The output of the rheometer is the typical torque-time relationship, where the y-axis indicates the torque in miliNewton-meter (mNm), and the x-axis indicates the duration of the rotational shear force applied to the sample (in seconds).

2.6 TTQ Determination

The area under the torque-time curve (TTQ) is required to determine the network strength in a torque rheogram. The software included with the rheometer allows a quick calculation of TTQ through setting slice lines included with the rheogram. There are four slice lines for analysis, two vertical and two horizontal (Figure 2-4). These slice lines can be located at any place on the rheogram. The location of the slice lines determines the boundaries of the data to be incorporated for analysis. Thus, it also determines the values for the peak of the rheogram and TTQ.

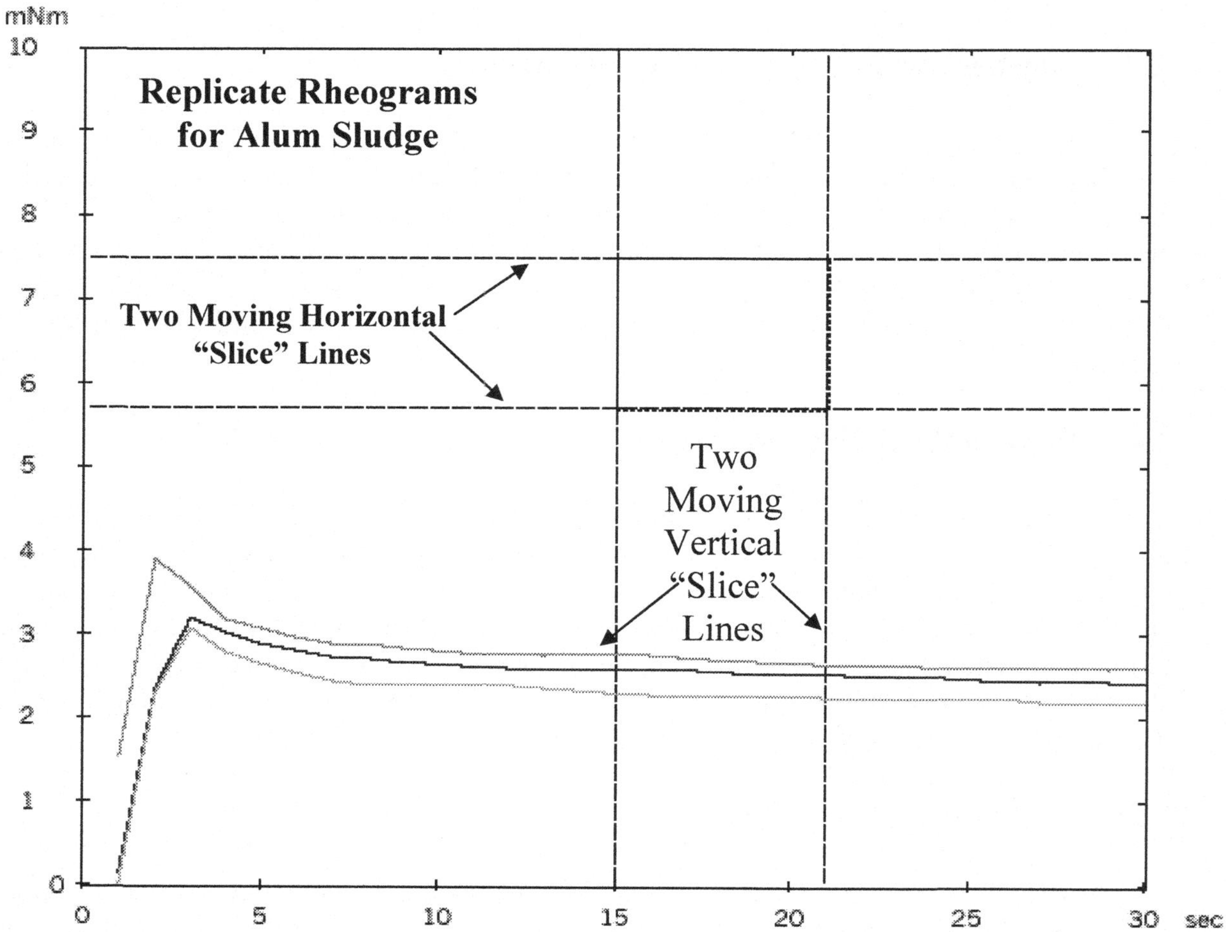

Figure 2-4. Typical Output of the Floccky Torque Rheometer Indicating the Four Slice Lines Used to Calculate the Area Under the Rheogram.

To determine the area under the rheogram, the bottom horizontal slice line should be set exactly at the x-axis, and the upper slice line should be set higher than the rheogram itself. This way the whole area under the rheogram remains within the boundaries of the horizontal slice lines, and the software is set to include all of the rheogram points in data analysis. For the vertical slice lines, the left vertical slice line should be set exactly at the y-axis, but the right vertical slice line can be set at different time positions, such as 5, 10, 15, 20, or 25 seconds. When the right vertical slice line is set at 15 seconds, for example, the software measures the area under the rheogram for up to 15 seconds.

2.7 Network Strength Determination

Area under the torque-time rheogram is used to estimate the relative network strength. The unit work energy is defined as the work energy required for processing a unit volume. This energy is representative of the energy dissipated into the suspension, which indicates the suspension's network strength, and can be calculated from equation 1.5.1-5 (see Chapter 1.0 for more details):

$$W_u = 2\pi N \times (TTQ)/V_b$$ where,

W_u = unit work energy and has units of (Joule/cm^3);
N = rotational speed in rpm;
TTQ = totalized torque, area under the rheogram and has units of (Joule-min); and
V_b = the processed volume samples in cm^3

The proposed method does not measure the "absolute" network strength, but a comparative strength of different aggregates using the same instrument and measuring conditions. Thus, the measured strength is relative, and it is cumulative of the network strength over time.

2.8 Full-scale Testing

Full-scale testing was conducted at the Wilmington Pollution Control Facility (Wilmington, Delaware) operated by Veolia Water North America Operating Services. Dose-response testing was conducted on both centrifuges (Bird Model 6150, designed for a max feed rate of 400 gpm) and gravity belt thickeners (Roediger Model 832.3, 3.2 meter width). The dose-response testing protocol as documented by Dentel et al. (1995) was used. Generally, the procedures forced incremental changes in the polymer dose, allowing the thickening or dewatering device to stabilize with the polymer feed, and requiring sampling of the output for appropriate measurements. Samples of the conditioned sludge with the different polymer doses were collected from both dewatering devices, and rheological measurements were conducted on site as described in Section 2-5.

CHAPTER 3.0

EXPERIMENTAL PHASES

The study featured laboratory and full-scale phases.

The laboratory bench-scale testing phase was designed to address the following issues:

- Compare the effect of impeller speed on rheological measurements.
- Establish the reproducibility of rheological measurements using different types of sludge.
- Determine the optimum rotational speed for different types of sludge.
- Determine how the optimum dose and rheograms change at different rotational speeds of the impeller.
- Determine the change in the rheograms at the optimum polymer dose.
- Determine the effect of mixing intensity and time used to condition the sludge on rheological measurements and optimum polymer dose.
- Determine the effect of polymer type on rheological measurements and optimum polymer dose.
- Determine the effect of inorganic conditioner ferric chloride on rheological measurements and optimum polymer dose.

The full-scale phase was designed to determine if the output of the rheometer obtained from bench-scale tests, expressed as network strength, could be used to determine full-scale optimum conditioning and dewatering. Testing occurred on a centrifuge and gravity belt thickener (GBT) at the Wilmington Pollution Control Facility. The polymer feed rate to the dewatering unit was changed incrementally, the unit was allowed to stabilize with the new polymer feed, and samples were collected for CST, total solids (TS), and rheology measurements.

CHAPTER 4.0

LABORATORY TESTS: RHEOLOGY REPRODUCIBILITY AND RELATION TO POLYMER DOSE

4.1 Reproducibility of the Rheological Measurements

As previously discussed, the reproducibility of the rheological measurements is central to this research. This section discusses the reproducibility of the rheograms for four residuals and the rheological measurements (torque, TTQ, network strength and peak torque) as a function of the rotational speed of the impeller. The rheology measurements were conducted at rotational speeds of 100, 150, 200, 250, 300, 350, 400, and 420 rpm. Very low speeds such as 100, 150, and 200 rpm resulted in zero or very low values of output signal, so these speeds were not considered further. Preliminary experiments showed that the impeller of the Floccky Tester could pick up changes in shear resistance at lower speeds when the solids concentration was higher than 1%.

4.1.1 Reproducibility of the Rheograms

To further investigate the reproducibility of the rheograms, measurements were repeated 20 times for different sludge types at impeller speeds of 300, 350, and 400 rpm. The TS content of the sludge types uses were: 2.17% for anaerobically digested sludge, 2.63% for aerobically digested sludge, 0.84% for activated sludge, and 1.56% for alum sludge. The rheograms for the sludge types at the three different rotational speeds is shown in Figures 4-1 to 4-4. The error bars indicate that good reproducibility is attained with the Floccky rheometer at all rotational speeds. Increasing the impeller speed, however, increased the magnitude of the error bars. Table 4-1 shows the average (± standard deviation [SD]) TTQ values after 120 seconds of shear application. Table 4-2 presents average (± SD) peak torque data. As the data indicate, results yielded good reproducibility of measurements.

Table 4-1. Average (±SD) for TTQ at 120 Seconds of 20 Samples for Different Sludge Types.

Speed	Anaerobic Sludge		Activated Sludge		Aerobic Sludge		Alum Sludge	
	Average	±SD	Average	±SD	Average	±SD	Average	±SD
300 rpm	245.96	±30.08	29.41	±22.84	181.13	±22.87	6.49	±2.60
350 rpm	524.34	±30.15	234.04	±32.02	451.75	±33.16	177.75	±45.51
400 rpm	703.38	±44.44	390.82	±42.35	655.49	±48.28	395.64	±39.98

Table 4-2. Average (±SD) for Peak Torque of 20 Samples for Different Sludge Types.

Speed	Anaerobic Sludge		Activated Sludge		Aerobic Sludge		Alum Sludge	
	Average	±SD	Average	±SD	Average	±SD	Average	±SD
300 rpm	2.85	±0.20	2.08	±0.29	2.48	±0.20	2.32	±0.77
350 rpm	4.57	±0.25	2.98	±0.18	3.98	±0.25	3.16	±0.76
400 rpm	6.10	±.40	3.62	±0.29	5.65	±0.42	3.96	±0.62

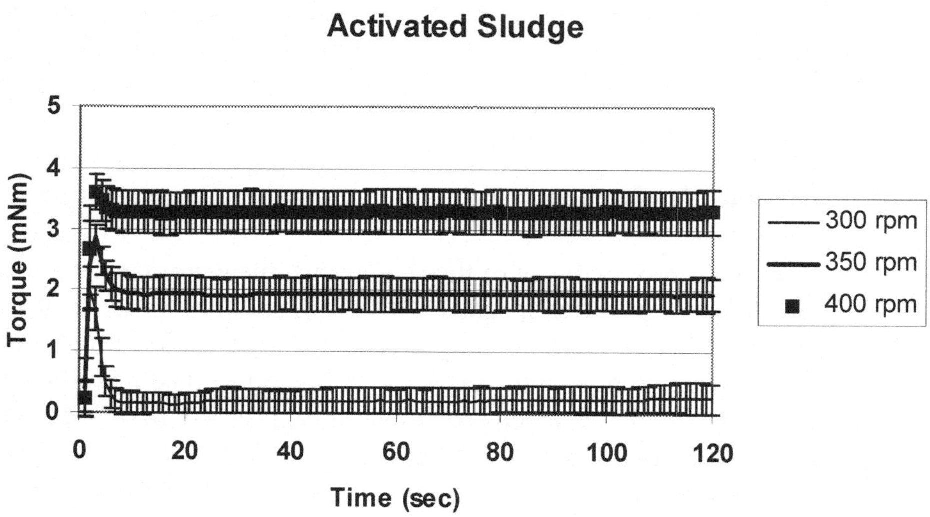

Figure 4-1. Averages (±SD) of Measurements at Different Impeller Speeds from 20 Activated Sludge Samples.

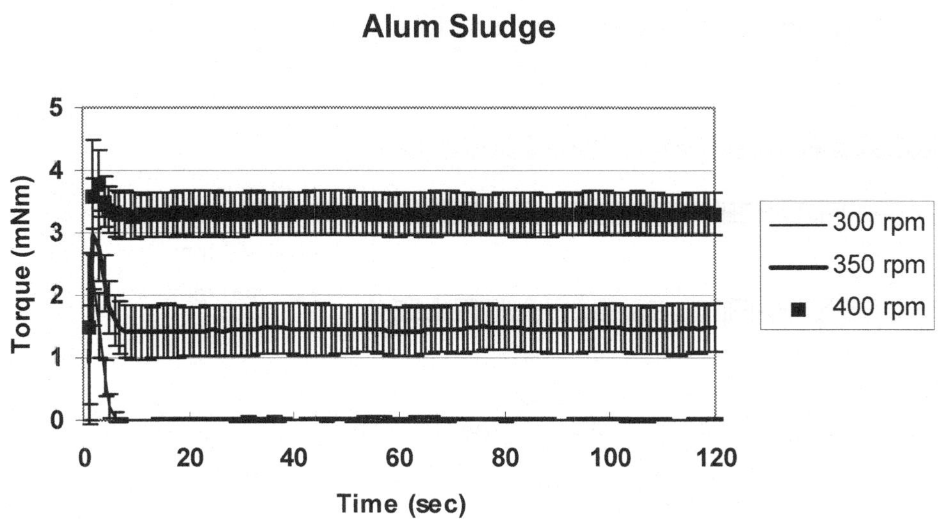

Figure 4-2. Averages (±SD) of Measurements at Different Impeller Speeds from 20 Alum Sludge Samples.

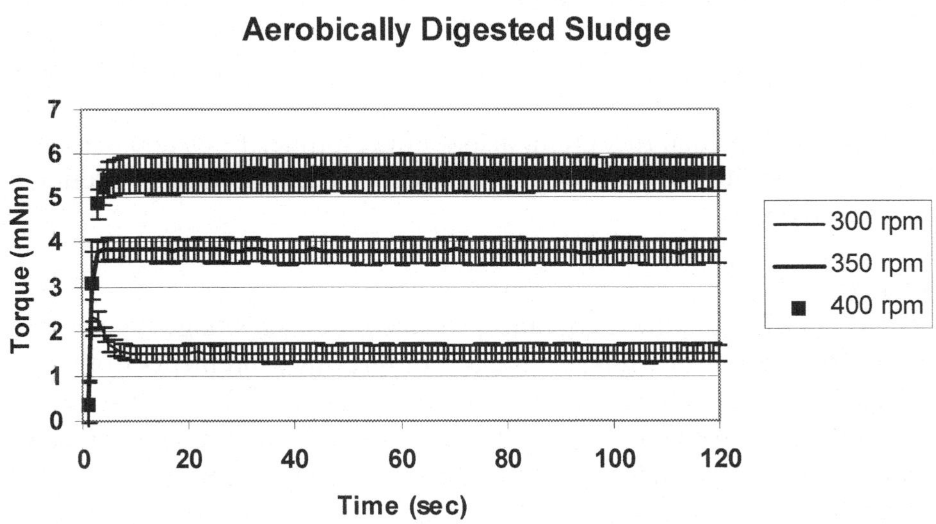

Figure 4-3. Averages (±SD) of Measurements at Different Impeller Speeds from 20 Aerobically Digested Sludge Samples.

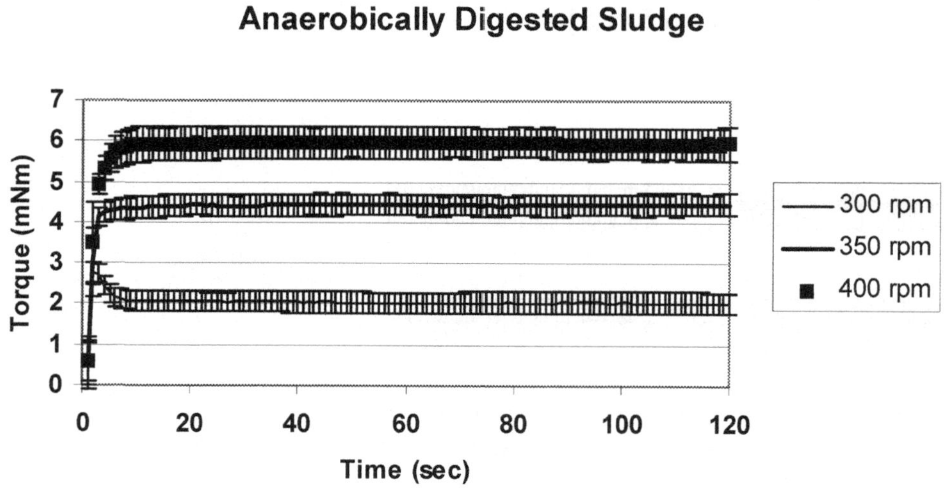

Figure 4-4. Averages (±SD) of Measurements at Different Impeller Speeds from 20 Anaerobically Digested Sludge Samples.

4.1.2 Effect of Rotational Impeller Speed on Rheological Parameters

This research examined the effect of rotational impeller speed and the relationship between the impeller speed and reproducibility of measurements on the rheological parameters of the four types of sludge. For each rheogram, peak torque, torque, TTQ, and network strength values were calculated at 20, 40, 60, 80, 100, and 120 seconds as described in Section 2.6 and 2.7 of Chapter 2, respectively.

4.1.2.1 Rheograms

Figures 4-5 to 4-8 illustrate the rheograms produced for each sludge type at different impeller speeds. Each data point shown is the average of three measurements from three sludge samples. Error bars are excluded to keep the figures clear. (The reproducibility of the rheograms and the magnitude of the error bars will be covered in detail in the next sections.) All of the sludge types, except the activated sludge, produced rheograms starting from an impeller speed of 250 rpm. Activated sludge, having a solids content of only 0.84%, did not produce a significant output signal at this speed, thus the slowest impeller speed used was 300 rpm. The reproducibility of rheograms were very good overall. The intensity of the output signal, or resistance to shear, increased with an increase in impeller speeds.

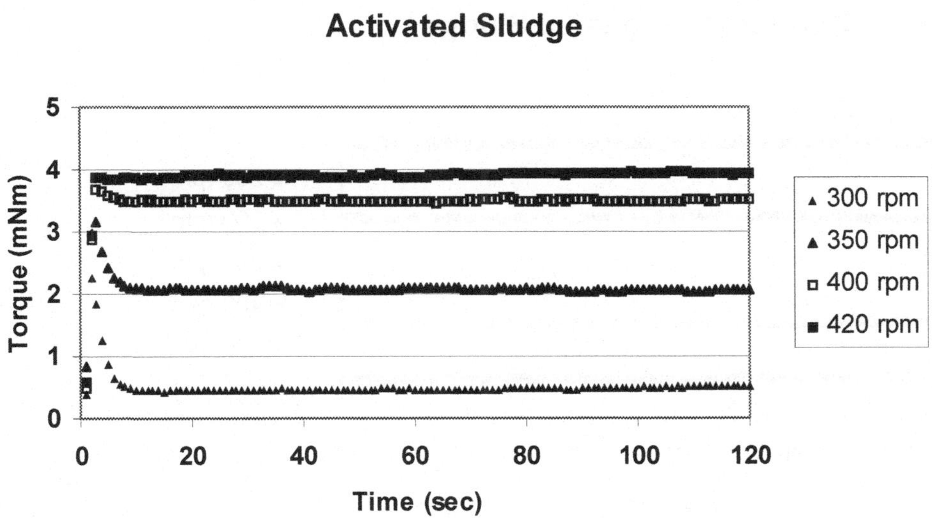

Figure 4-5. Rheograms for Activated Sludge at Different Impeller Speeds.

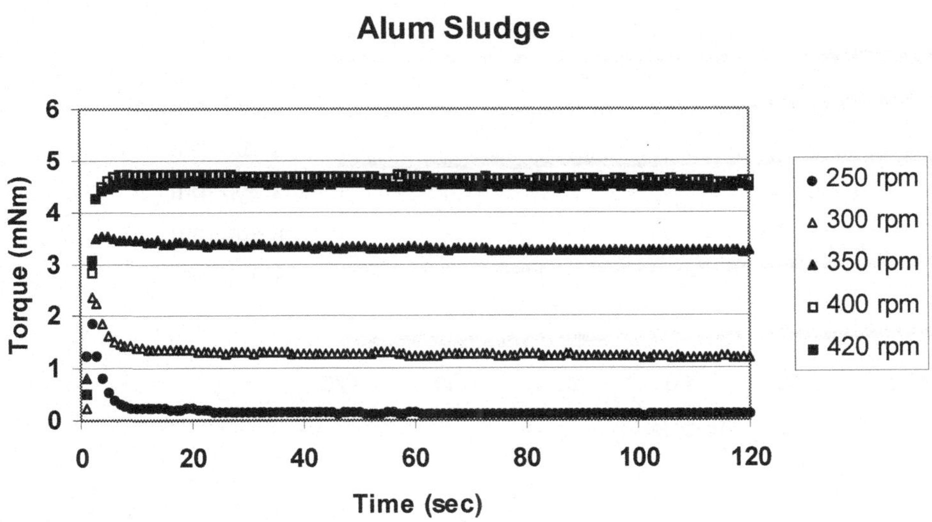

Figure 4-6. Rheograms for Alum Sludge at Different Impeller Speeds.

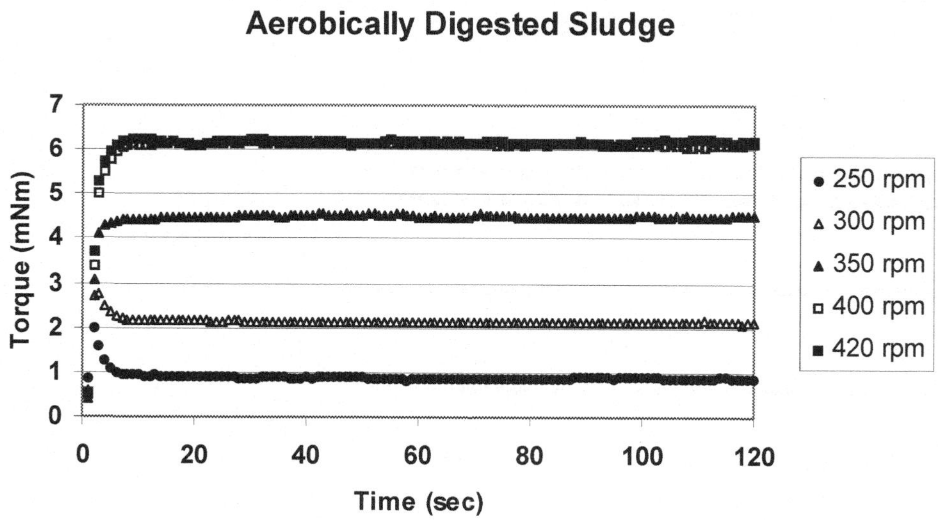

Figure 4-7. Rheograms for Aerobically Digested Sludge at Different Impeller Speeds.

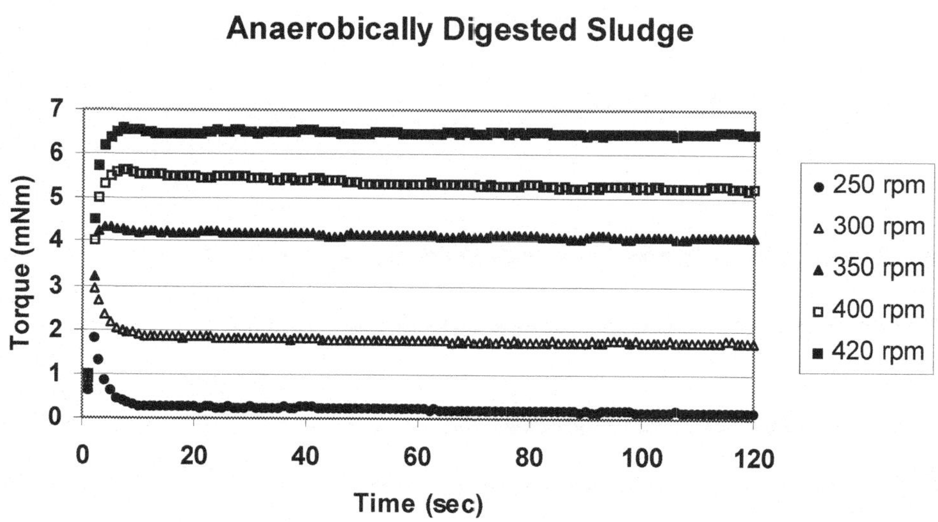

Figure 4-8. Rheograms for Aanaerobically Digested Sludge at Different Impeller Speeds.

4.1.2.2 TTQ

The TTQ was determined from the Floccky software at 20, 40, 60, 80, 100, and 120 seconds. Figures 4-9 to 4-12 show the calculated TTQ for the four different sludge types at the chosen times. There was a linear relationship between the area under the rheogram (TTQ) and the time of applying the shear at all of the impeller speeds. In addition, higher impeller speeds yielded higher TTQ values.

Figure 4-13 illustrates the calculated TTQ values at each impeller speed at 120 seconds after the start of the measurement. TTQ increased with increasing impeller speeds. In addition, TTQ values were related to the solids content of the sludge samples. When the solids content increased, TTQ increased.

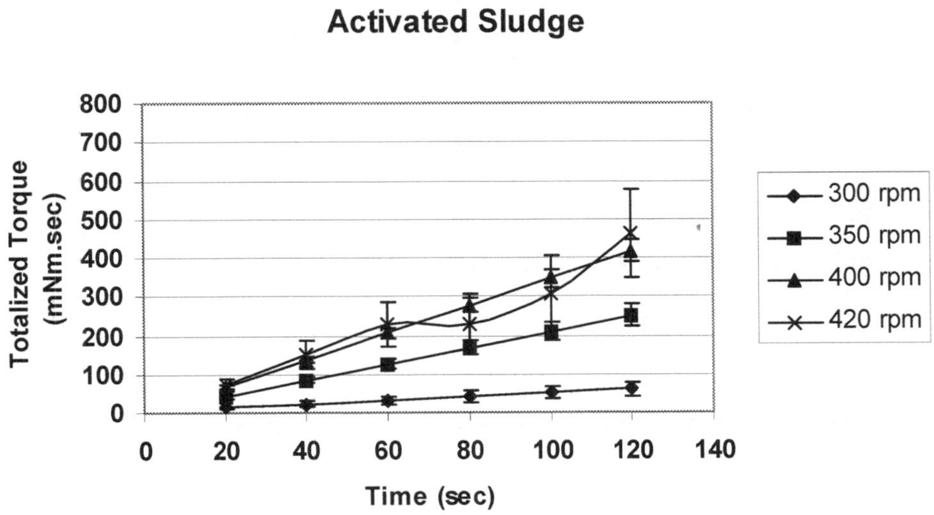

Figure 4-9. TTQ Values for Activated Sludge at Different Impeller Speeds and Times of Shear.

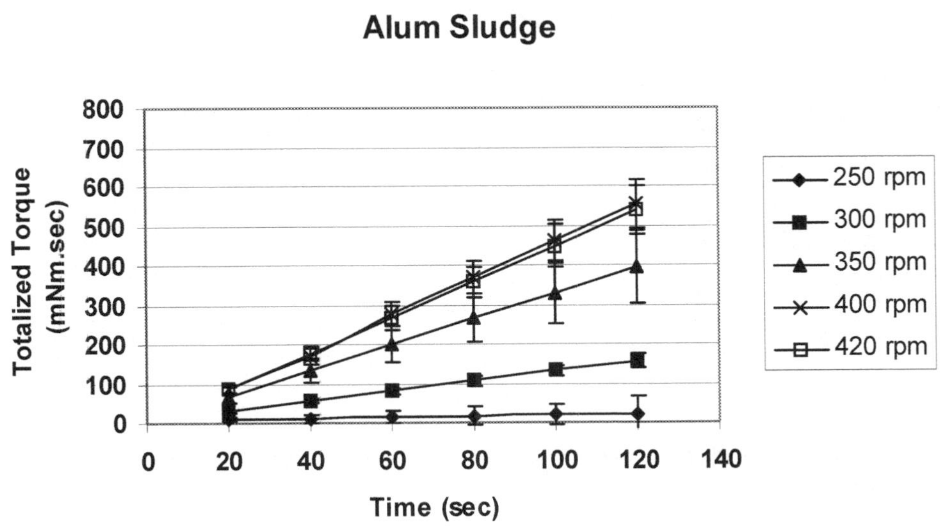

Figure 4-10. TTQ Values for Alum Sludge at Different Impeller Speeds and Times of Shear.

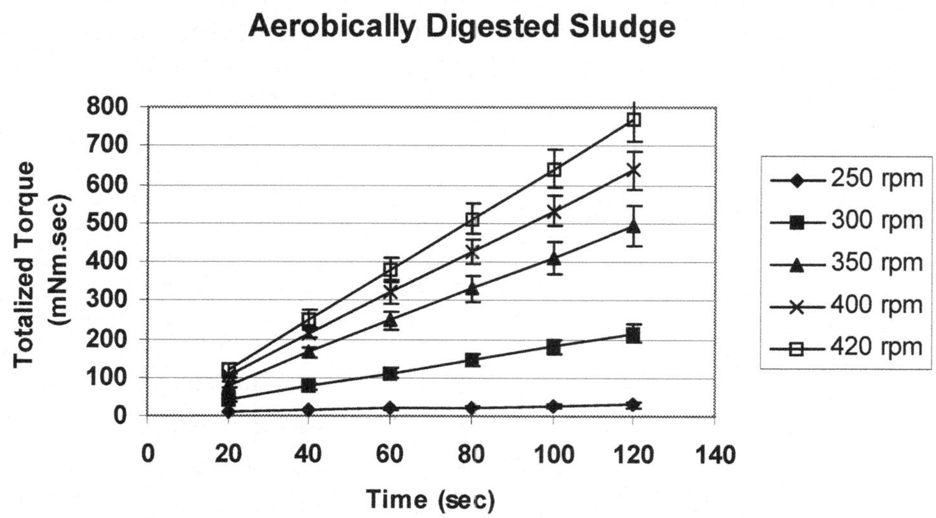

Figure 4-11. TTQ Values for Aerobically Digested Sludge at Different Impeller Speeds and Times of Shear.

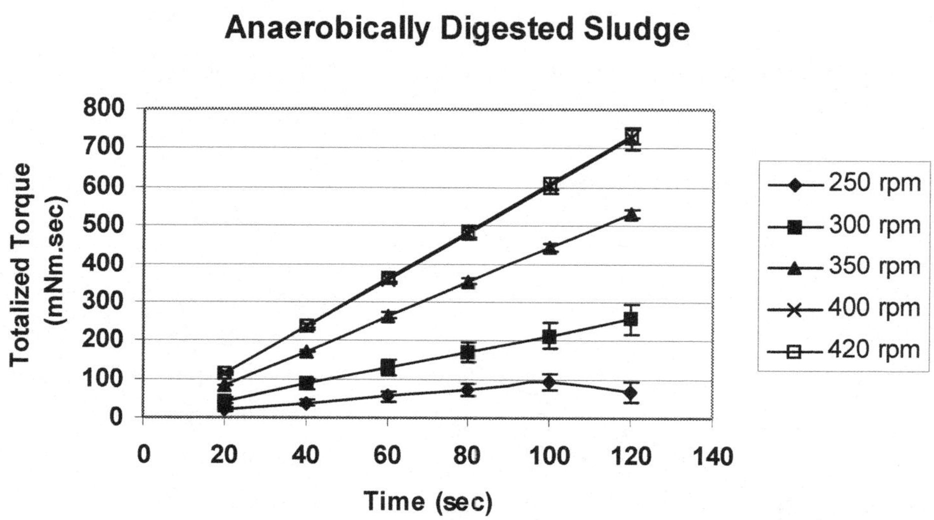

Figure 4-12. TTQ Values for Anaerobically Digested Sludge at Different Impeller Speeds and Times of Shear.

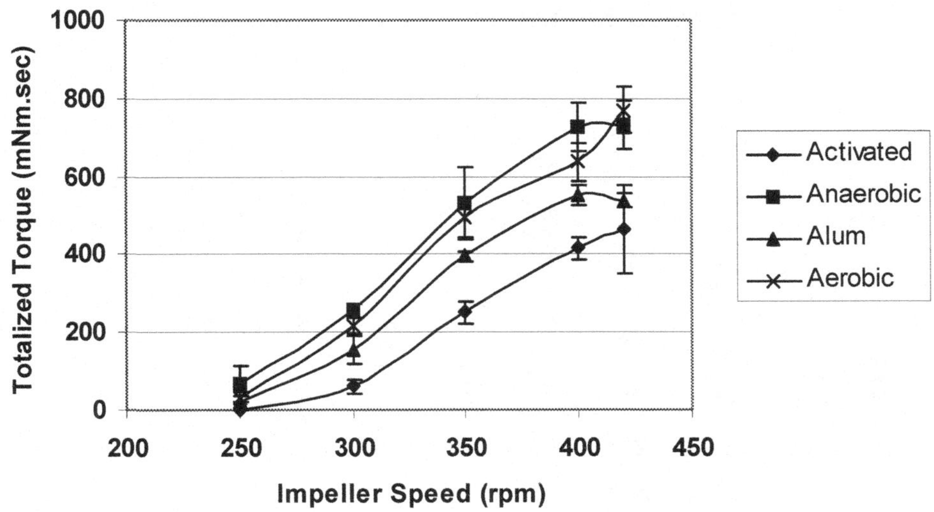

Figure 4-13. TTQ Values for Activated, Alum, Aerobically Digested, and Anaerobically Digested Sludge at Different Impeller Speeds 120 Seconds After the Start of the Measurement.

4.1.2.3 Network Strength

Network strength was calculated by incorporating the TTQ values in the equation provided in Chapter 2.0. Figures 4-14 to 4-17 illustrate the network strength the sludge types at different impeller speeds. Similar to TTQ, network strength also has a linear relationship with the time of shear application and the impeller speed.

Comparison of the network strengths of four of the sludges revealed that activated sludge had the highest network strength even though it produced the lowest TTQ (Figure 4-18). The high network strength of activated sludge may be explained by the high concentration of extracellular polymers present in activated sludge, which help to hold sludge flocs together.

The peak torque values observed for each sludge type are illustrated in Figure 4-19. Peak values also seemed to increase with increasing impeller speeds.

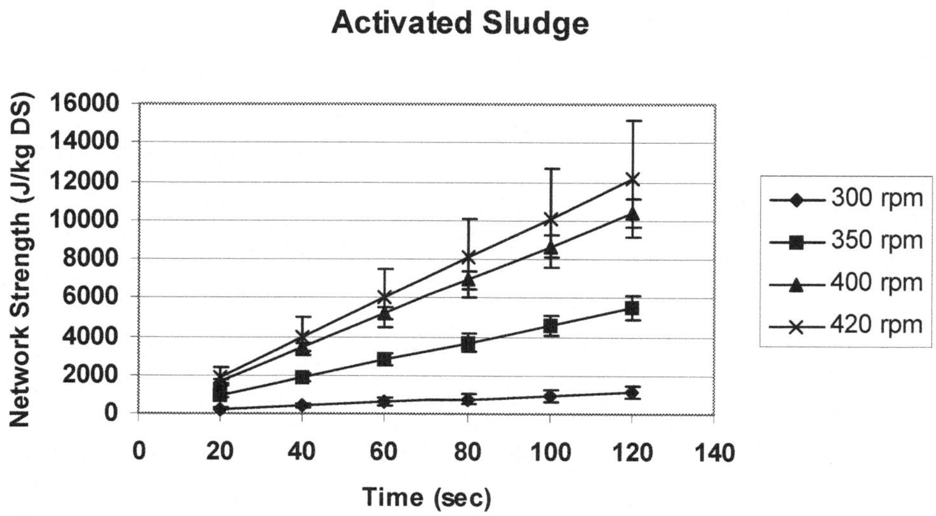

Figure 4-14. Network Strength Values for Activated Sludge at Different Impeller Speeds and Times of Shear.

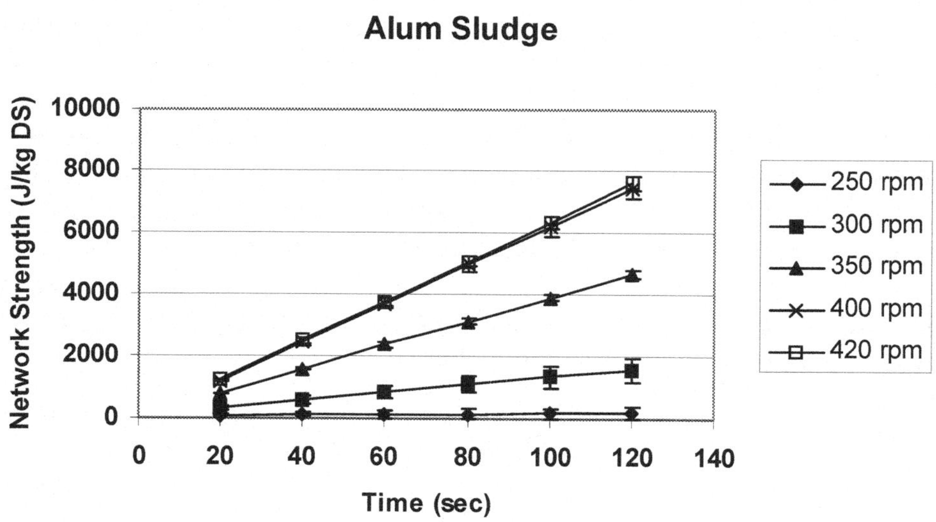

Figure 4-15. Network Strength Values for Alum Sludge at Different Impeller Speeds and Times of Shear.

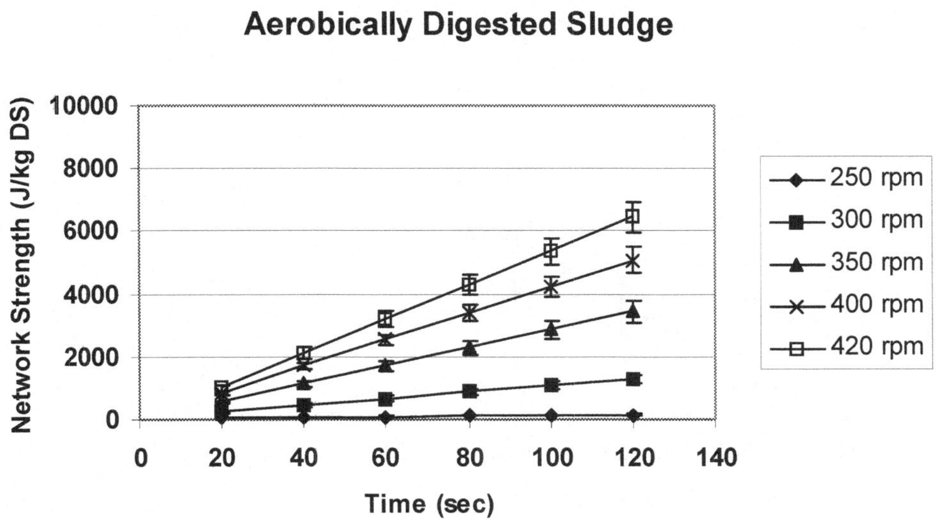

Figure 4-16. Network Strength Values for Aerobically Digested Sludge at Different Impeller Speeds and Times of Shear.

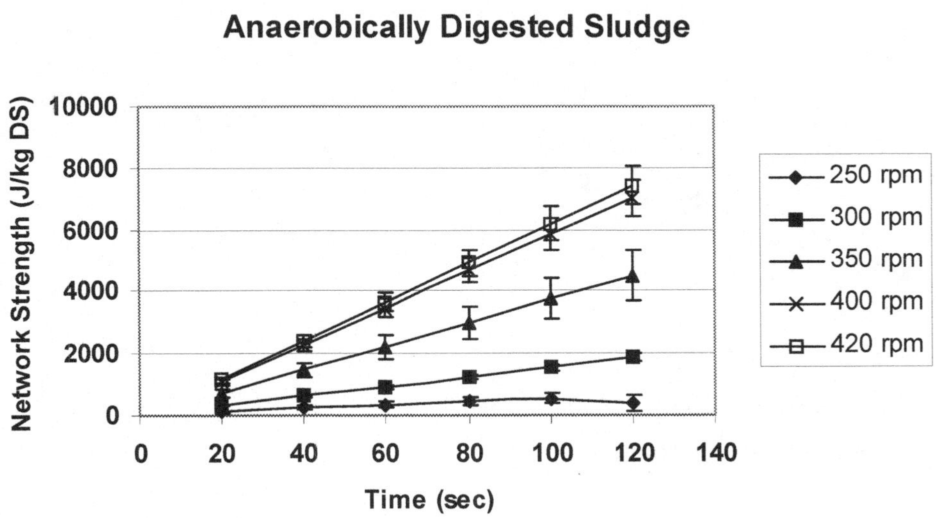

Figure 4-17. Network Strength Values for Anaerobically Digested Sludge at Different Impeller Speeds and Times of Shear.

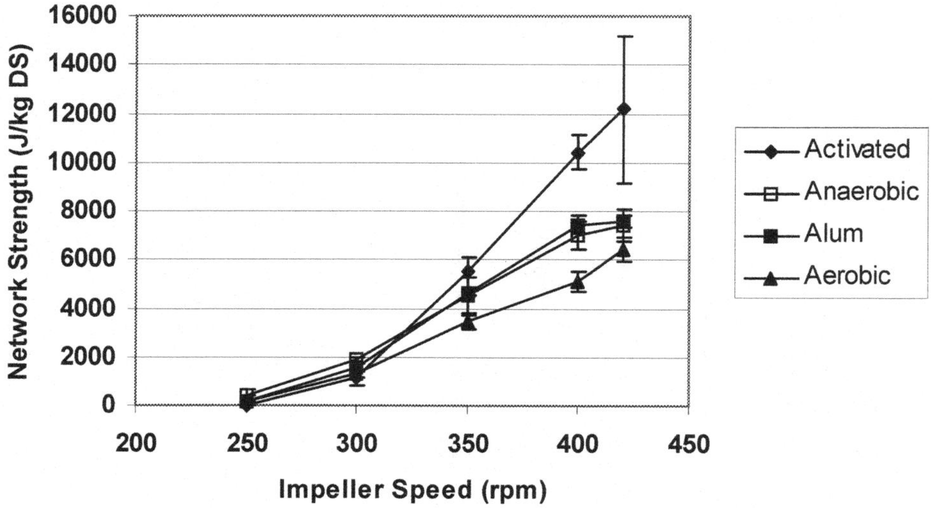

Figure 4-18. Network Strength Values for Activated, Alum, Aerobically Digested, and Anaerobically Digested Sludge at Different Impeller Speeds.

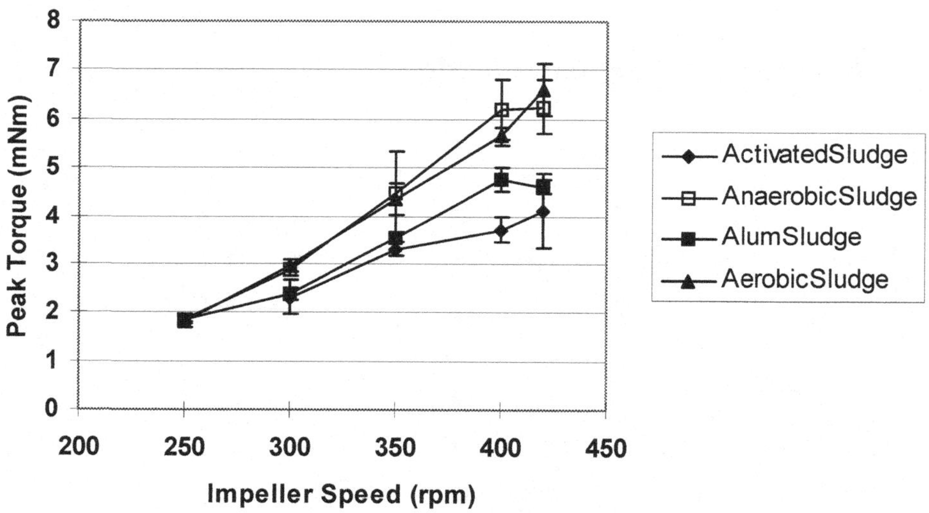

Figure 4-19. Peak Torque Values for Activated, Alum, Aerobically Digested, and Anaerobically Digested Sludge at Different Impeller Speeds.

4.2 Correlation of Rheological Parameters to Polymer Dose

One of the main objectives of this research was to correlate the torque rheology parameters directly to polymer dose. Different polymer doses were used with mixing conditions of 200 rpm for 120 seconds. The experiments reported in this section used anaerobically digested sludge from the North Durham Water Reclamation Facility. The polymer (Stockhousen 655K) as practiced at the plant was used for the conditioning experiments at a 0.5% solution. The polymer dose was increased in a step-wise manner. Various doses of polymer were added to 200 mL aliquot samples of the same batch of sludge collected for these experiments. Rheological measurements were conducted in triplicate as described in Chapter 2.0. In addition, CST and filtration tests were carried out to measure the dewaterability of the sludge types as described in Chapter 2.0. The same measurements were repeated for each polymer dose.

The effect of different types of polymers on rheological properties was also studied and is discussed in Chapter 5.0. The results presented in this section describe the performance of Stockhausen 655K.

4.2.1 Dewatering Results

The CST (Figure 4-20) was the lowest at the 13.6 g/kg dry and 17 g/kg dry solids (DS) polymer dose. A high degree of flocculation was observed where representative sub-samples of suspension were not feasible for CST measurements, indicating polymer overdosing. Thus, the CST value at 17 g/kg DS (the highest dose) is not reported.

Figure 4-21 presents the measured filtrate volume after 10 minutes of filtration time of the 200 mL flocculated sample. At optimum dewatering, the amount of water that was withdrawn from the sludge is the highest. Thus, the optimum polymer conditioning occurred at the 13.6 g/kg DS dose. This dose should also correspond to the cake with the highest solids content that can be obtained following filtration. This is evident from the same figure that shows a maximum cake of about 7.5% solids at 13.6 g/kg DS sludge.

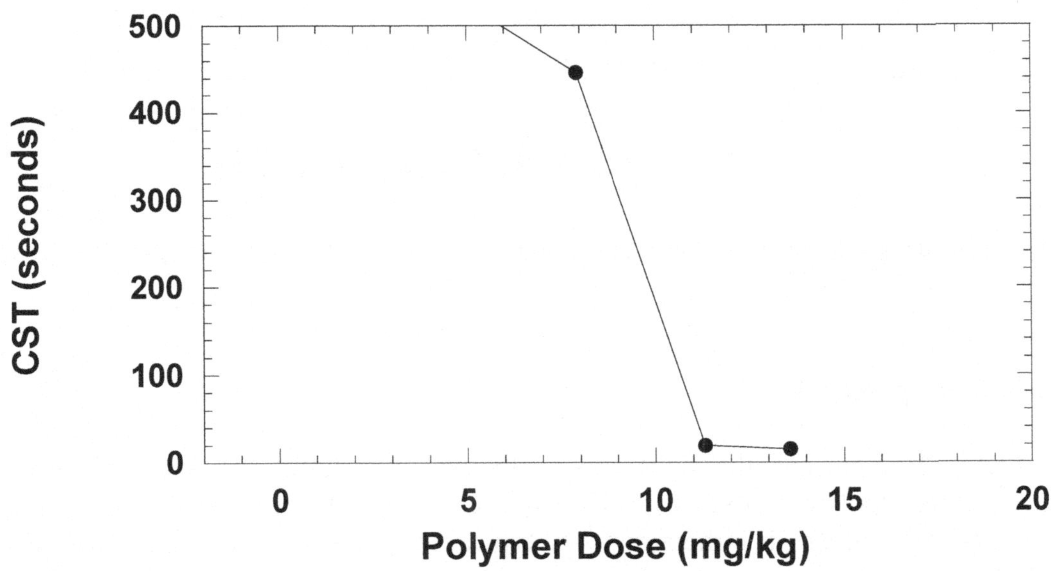

Figure 4-20. CST of Anaerobically Digested Sludge Samples at Different Polymer Doses.

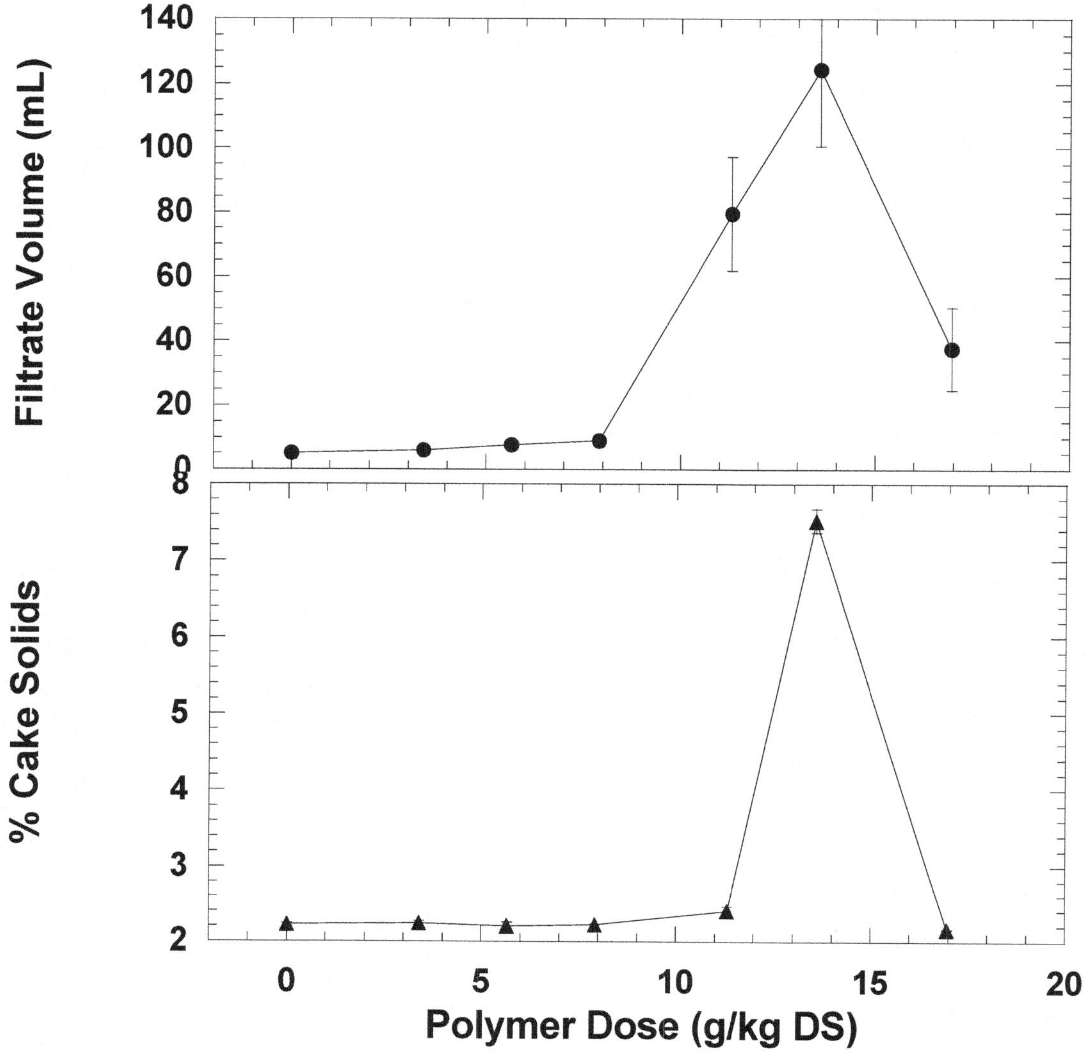

Figure 4-21. Filtrate Volume and Percent Cake Solids of Anaerobically Digested Sludge Samples After 10 Minutes of Filtration.

4.2.2 Rheology Results

Figure 4-22 presents the rheograms for the raw and conditioned anaerobically digested sludge at different polymer doses as previously used in the filtration results. Three replicate rheograms were obtained. The rheograms are typical torque vs. shearing time relationships. In general, the rheogram replicates for the same polymer doses show good reproducibility at lower polymer doses of up to 11.3 g/kg DS.

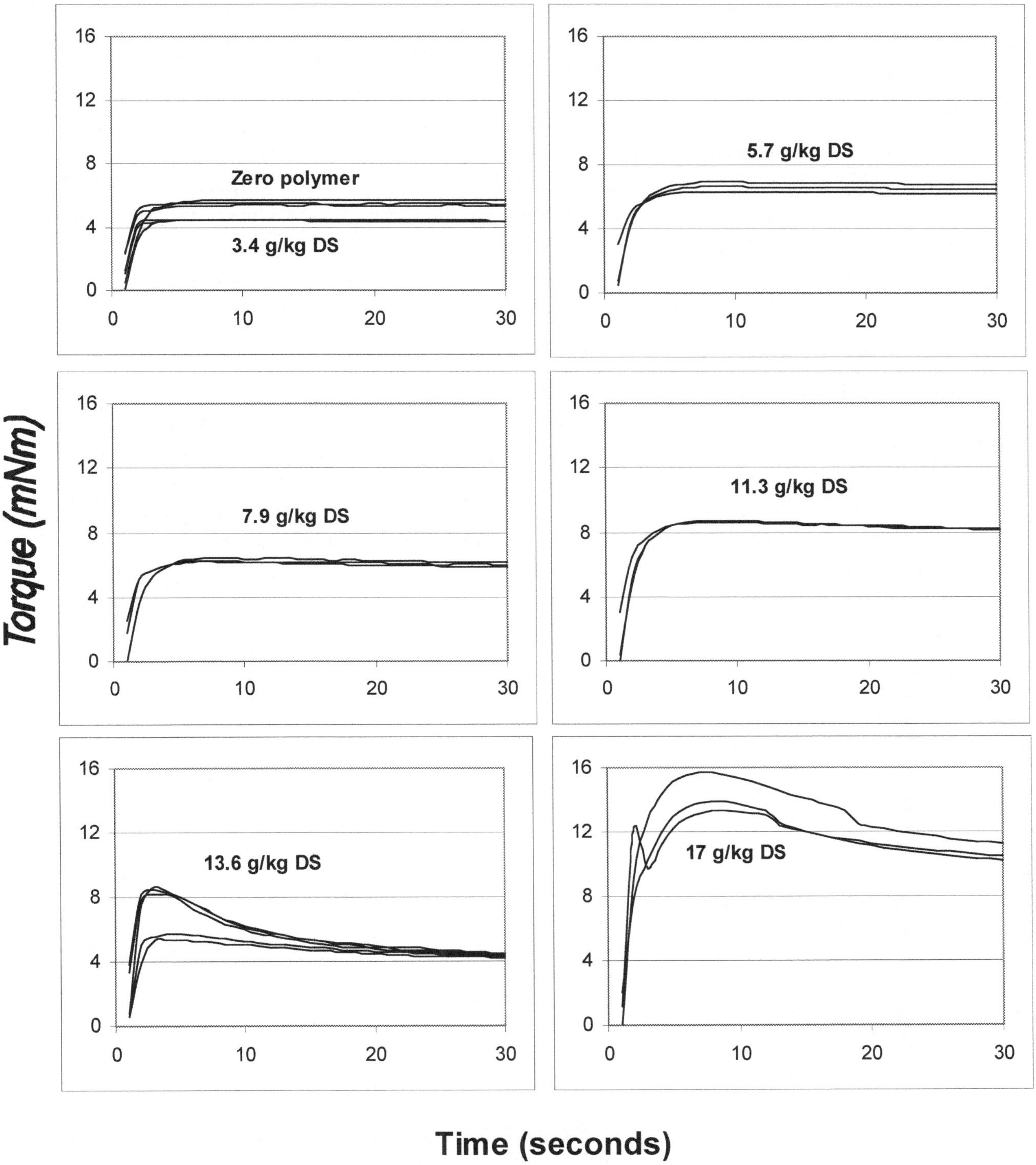

Figure 4-22. Rheograms from Unconditioned and Conditioned Anaerobically Digested Sludge Samples.

Three important observations can be made from these rheograms. First, after conditioning with the lowest polymer dose of 3.4 g/kg DS, the measured torque decreased when compared to the original sludge without conditioning. The only explanation that we can offer is that this low polymer dose led to deterioration in the sludge structure. Second, the three rheograms at 13.6 g/kg DS exhibited a more evident torque peak as compared to the rheograms representing the lower polymer doses. Moreover, the torque values at the higher shearing times for 13.6 g/kg DS were lower than observed in the lower polymer doses. This dose represents the optimum dose for dewaterability as evident from the filtration (Figure 4-21) and CST (Figure 4-20). The torque measurements for this dose showed variations in the rheograms produced on five replicates. The final observation can be made from the highest polymer dose of 17 g/kg DS, where the measured torque increased drastically. One rheogram indicated floc breakage (first peak) at the earlier stage of shearing. However, due to the excess polymer in the suspension, it is likely that re-flocculation occurs and results in the formation of the second peak.

4.2.2.1 Rheological Data Analysis—Peak Torque

Although rheogram figures (at the lower polymer dose) show no evident peak in the lower polymer doses, peaks were observed when the torque range was narrowed. Figure 4-23 shows the peak heights at the different polymer doses. Peak torque values are higher when more polymer is added to the sludge. Langer et al. (1994) observed the same phenomena. Note that the peak height does not indicate the polymer optimum dose and that the higher doses caused large variations in the peak values, and thus reproducibility of the results.

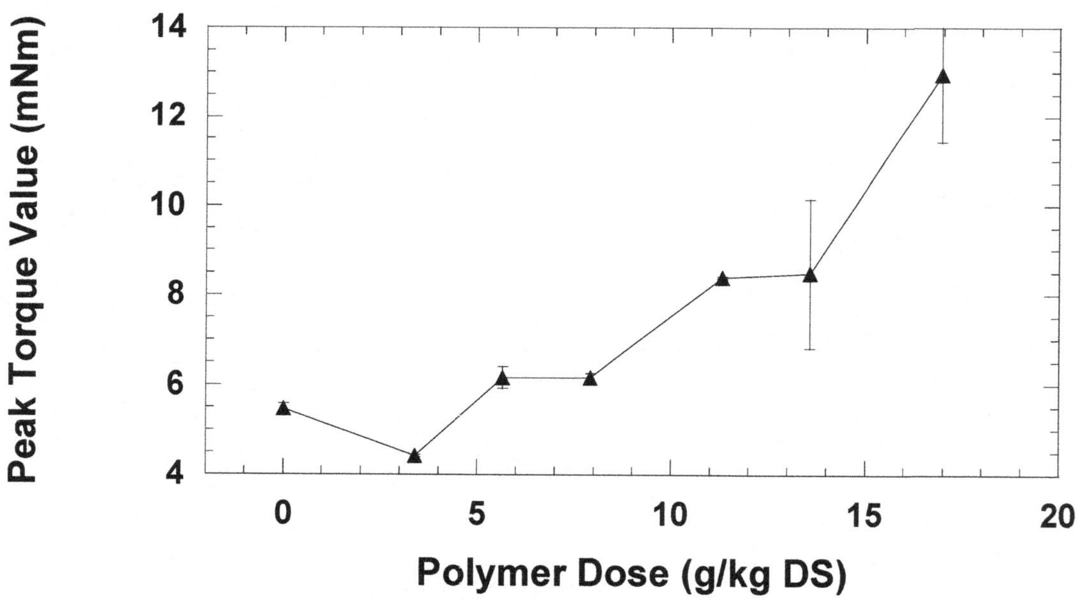

Figure 4-23. Peak Torque Values as a Function of the Polymer Doses Presented in Figure 4-21.

4.2.2.2 Rheological Data Analysis—Network Strength

Figure 4-24 presents the network strength values calculated as previously described in Chapter 2.0. Three replicate measurements made at different shearing times (5, 10, 15, 20, 25, 30 seconds) are plotted against different polymer doses. At a specific shear time, increasing the polymer dose lead to an increase in the network strength up to a certain dose ~13.6 g/kg DS, at which the network strength drops. Increasing the polymer dose above this dose lead to an increase in the network strength again.

The lower network strength at the 13.6 g/kg DS polymer dose allowed the sludge to release most of its incorporated water while maintaining its structure. This is evident from the filtration results for the same samples where the highest filtrate volume and cake solids content occurred at this dose where the energy suddenly decreased to lower values (Figure 4-21). The release of water led to good suspension dewaterability without complete floc deterioration while some type of shear resistance was maintained. At the higher polymer doses, residual, unflocculated polymer in the system may re-flocculate the broken network bonds, leading to higher energy levels.

Figure 4-24 also shows the torques at different shear times by polymer dose. The trend in measured torque values matches the trends observed for the energy presented as network strength because of their relationship as described in Chapter 2.0. Thus, the torque itself can be used as an indication of the network strength of the sludge at different polymer doses since torque must be considered after the observed peak and at an appropriate shear time.

The energy required to break the network bonds, which is important in terms of dewaterability for the polymer conditioned sludge, should be observed at higher shearing times. The energy contained in the network bonds allowed the polymer-conditioned material to withstand the shear that the sludge might experience in the dewatering device.

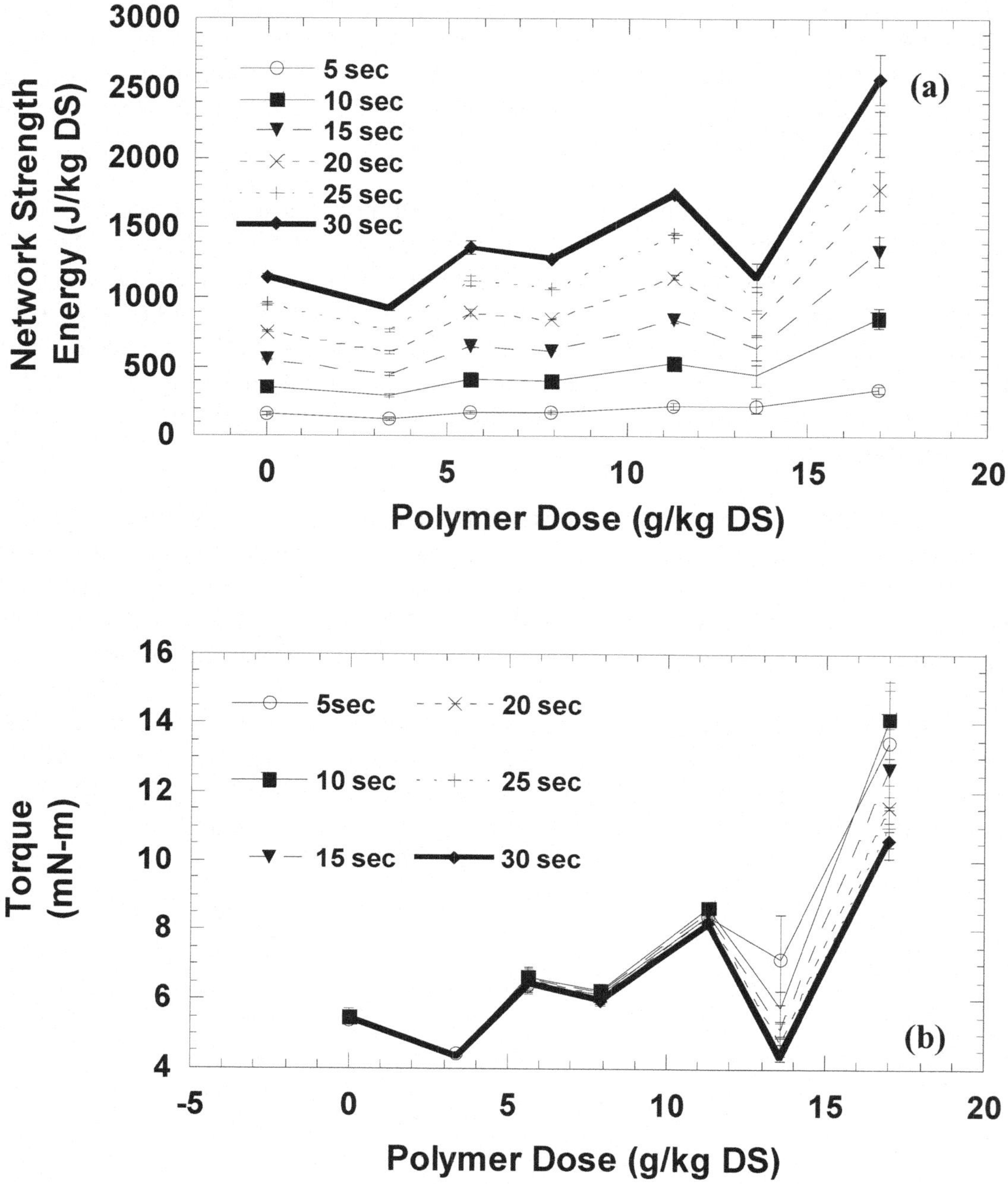

Figure 4-24. Network Strength and Torque Values at Different Times of Applying the Shear as a Function of Polymer Dose.

CHAPTER 5.0

LABORATORY TESTS: FACTORS AFFECTING MEASURED NETWORK STRENGTH

5.1 Effect of Impeller Speed on Network Strength and Optimum Dose

The objective of these experiments was to determine the relationship between the optimum polymer dose and change in the rheograms as a function of the impeller speed using anaerobically digested sludge.

Sludge samples were conditioned using a solution of 0.5% Stockhausen 655K polymer. After the addition of the polymer, samples were mixed at 300 rpm for 120 seconds using a jar tester. At least six and up to eight different polymer doses were tested to determine the optimum polymer dose. We took rheological measurements in triplicate, and calculated the torque, peak torque, and network strength. In addition, we determined CST, filtration volume, and percent solids of the cake after 10 minutes of vacuum-filtration at each polymer dose. The experiments were performed at impeller speeds of 300 rpm, 350 rpm, and 400 rpm.

5.1.1 Rheological Measurement Results

Figures 5-1, 5-2, and 5-3 show the rheograms of anaerobically digested sludge produced at a range of selected polymer doses at 300, 350, and 400 rpm respectively. The rheograms are the average of three measurements. Polymer doses of 0, 2, 4, 6, 8, and 10 g/kg DS were tested at all of the impeller speeds. In addition, polymer doses of 5 and 9 g/kg DS were tested when needed. (Not all of the polymer doses are presented in the figures to avoid cluttering of the figures.) Thus, rheogram figures illustrate just enough of the polymer doses to show the trend of the data.

Important conclusions can be drawn from Figures 5-1 through 5-3. First, the torque values measured by the instrument were lowest at 0 g/kg DS. Second, increasing the polymer dose increased the torque values. Torque values are an indication of resistance to shear, and at higher polymer doses resistance to shear will be higher due to increased network strength. Third, even though torque values increased steadily with increasing polymer doses, at a particular polymer dose torque values start to decrease and then continue to increase again at higher polymer doses. This decrease or "dip" on the rheograms can be seen at all of the impeller speeds tested.

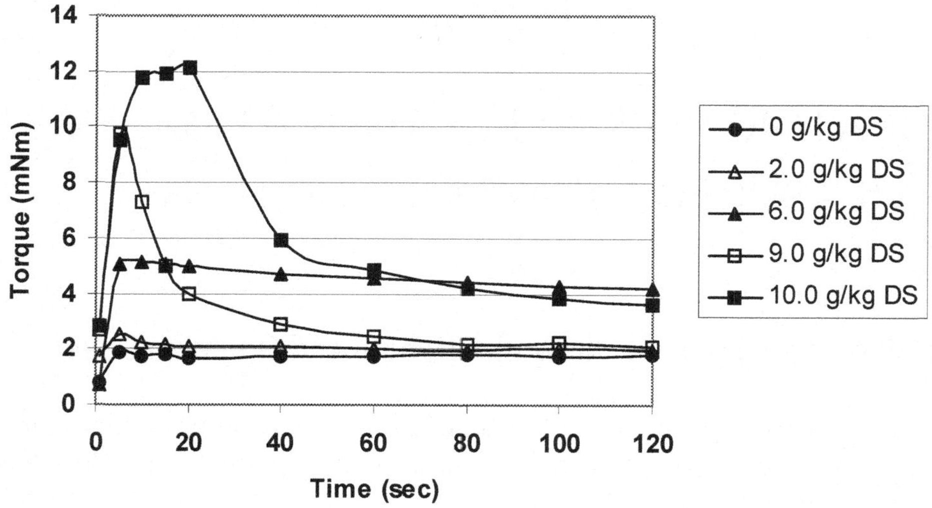

Figure 5-1. Rheograms Measured at 300 rpm Impeller Speed at Different Polymer Doses.

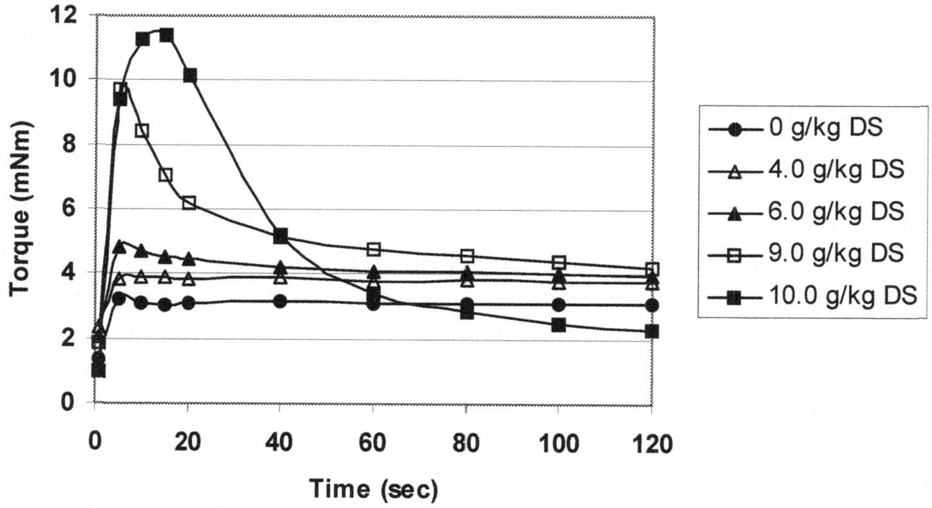

Figure 5-2. Rheograms Measured at 350 rpm Impeller Speed at Different Polymer Doses.

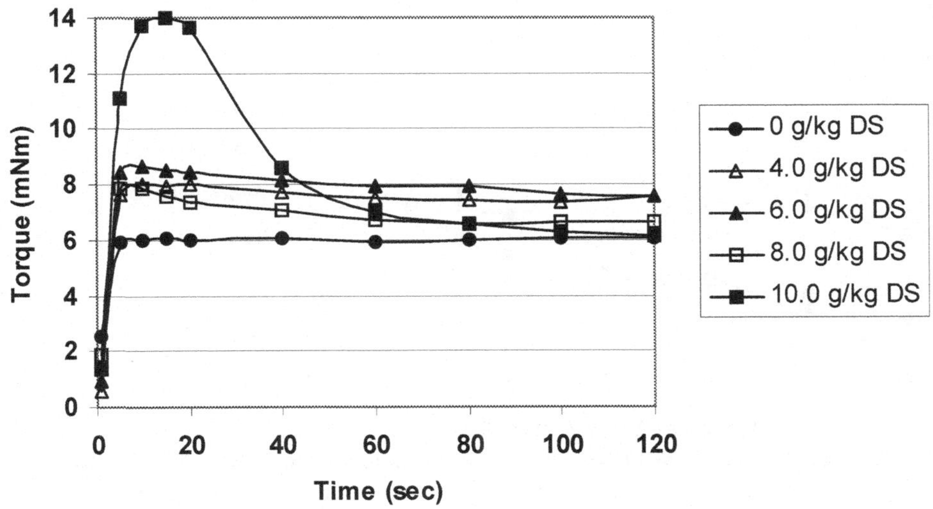

Figure 5-3. Rheograms Measured at 400 rpm Impeller Speed at Different Polymer Doses.

The information obtained from figures 5-1 through 5-3 can also be plotted as shown in Figure 5-4, which shows the torque values measured at 40 seconds at all of the polymer doses, at impeller speeds of 300, 350, and 400 rpm. In this figure, each polymer dose is represented by only one data point. The 40-second time was chosen because the majority of the peaks were formed within the first 5 to 10 seconds were over by 20 seconds, and the rheograms remained asymptotic at all polymer does and shearing times. A close examination of the previous rheograms also revealed that the tails of the rheograms started to converge at extended operation times such as 100 and 120 seconds. This reduces the difference that could be observed between the rheograms at higher shear times. *Thus, it is best to choose a time point that is after the formation of the peaks and before the convergence of the tails to compare torque values between different samples.* Our experience showed that this time point corresponded to 40 seconds for the majority of the rheograms. The exceptions were the overdosed samples in which the excess polymer in sludge caused the formation of large peaks that extended beyond 20 to 30 seconds. In the following sections of this report, the information obtained from the rheograms will be plotted using this method. This method is also preferred to avoid the visual clutter associated with plotting six to eight rheograms on one figure.

Figure 5-4 shows the trend of the rheograms and the dips observed around polymer doses of 8 to 9 g/kg DS. A decrease in the torque values starting at 8 g/kg DS can be seen at impeller speeds of 300 and 400 rpm. However, this dip is not observed at 350 rpm. This could be due to missing the polymer dose where the dip would have been observed during the rheological measurements (such as 7.5 or 8.5 g/kg DS), or an experimental error. At 10 g/kg DS, a decrease in the torque values was observed (Figure 5-2), however this probably is already the overdose region, and the critical polymer dose was missed in this experiment.

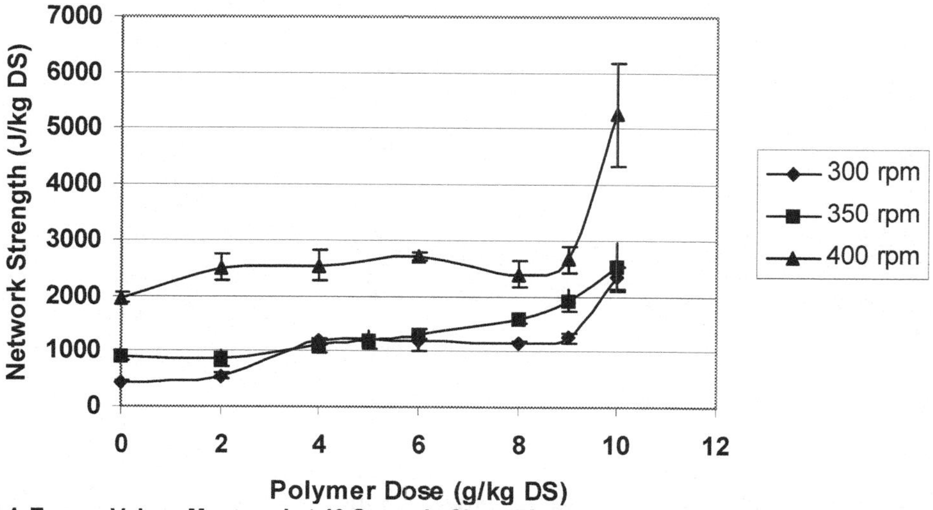

Figure 5-4. Torque Values Measured at 40 Seconds Shear Time at Different Impeller Speeds and Polymer Doses.

Figure 5-5 illustrates the network strengths calculated at 40 seconds for different polymer doses and at three impeller speeds. The rapid increase in the network strength beyond 9 g/kg DS indicates the overdose region. When the polymer dose is higher than the optimum dose, excess polymer in sludge results in the formation of large peaks which in turn increase the area under the rheogram considerably. Therefore, a rapid increase in the network strength is observed when polymer is overdosed. At 8 g/kg DS, network strength reaches its lowest values at impeller speeds of 300 and 400 rpm. This can be seen when scale of the y-axis is reduced to 3000 J/kg DS. The decrease in the network strength is not observed at 350 rpm possibly due to the reasons explained above.

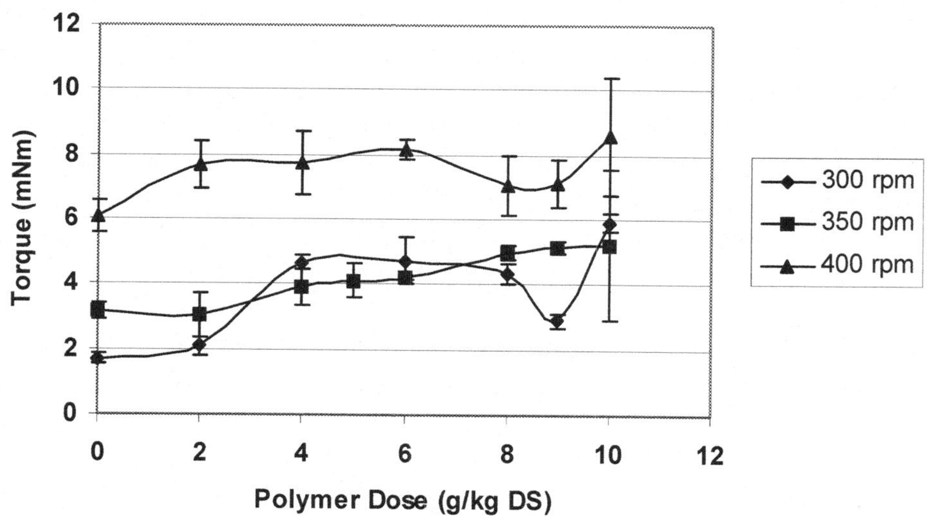

Figure 5-5. Network Strength Values Calculated at 40 Seconds Shear Time at Different Impeller Speeds and Polymer Doses.

The peaks of the rheograms follow a similar trend with those of the torque and network strength figures (Figure 5-6). Beyond 9 g/kg DS, a rapid increase in the peak values are observed. The dip at 8 g/kg DS is clear at both 300 rpm and 400 rpm.

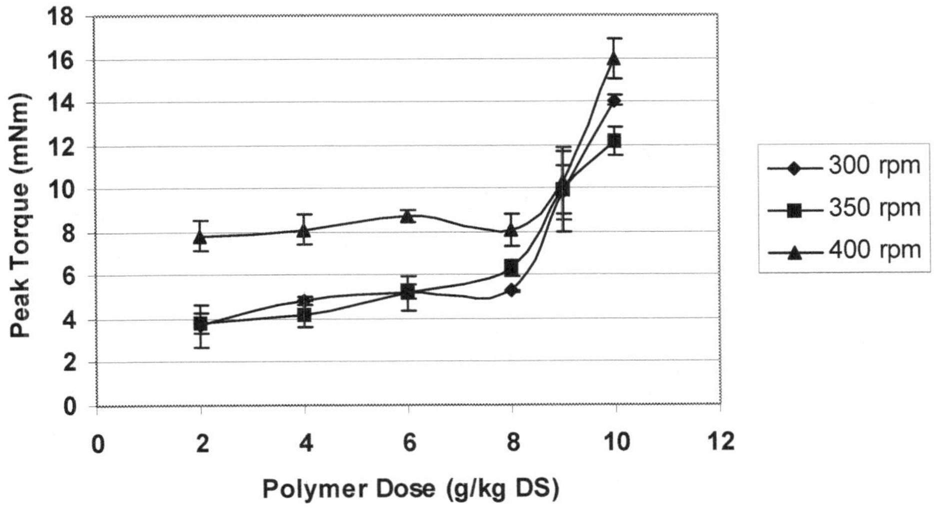

Figure 5-6. Peak Torque Values at Different Impeller Speeds and Polymer Doses.

5.1.2 Dewaterability Test Results

Figures 5-7 through 5-9 illustrate results from the dewaterability tests. CST reaches its lowest and the filtrate volume and percent cake solids are at their highest values at 8 g/kg DS. These results show that the optimum polymer dose for this sludge is 8 g/kg DS.

Results from these experiments indicate that a "dip" on the rheograms, or a decrease followed by an increase in measured torque, peak, and network strength values is observed at around the optimum dose. Thus, monitoring the changes in torque values of a sludge conditioned with various polymer doses is a useful tool to determine the optimum dose during sludge dewatering. An important observation is that the dip occurred at the optimum is observed at all of the impeller speeds tested.

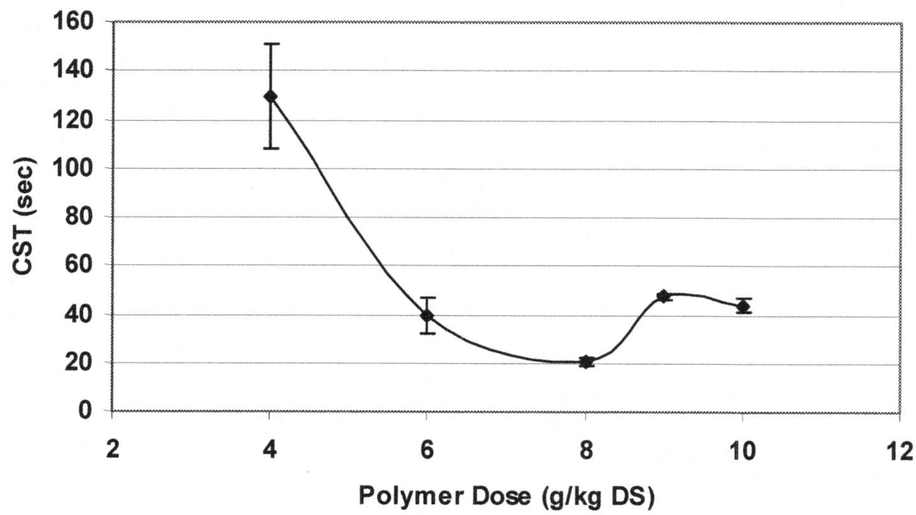

Figure 5-7. CST of Sludge at Different Polymer Doses.

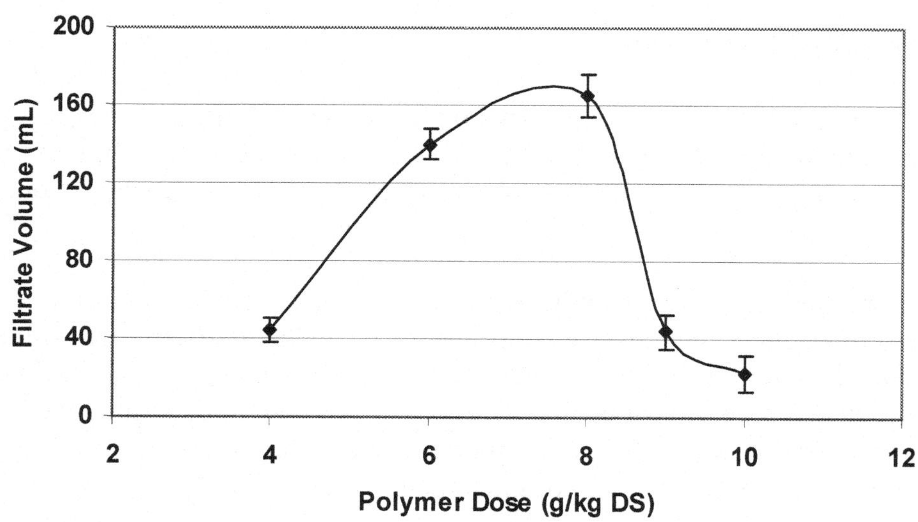

Figure 5-8. Filtrate Volume of Sludge at Different Polymer Doses.

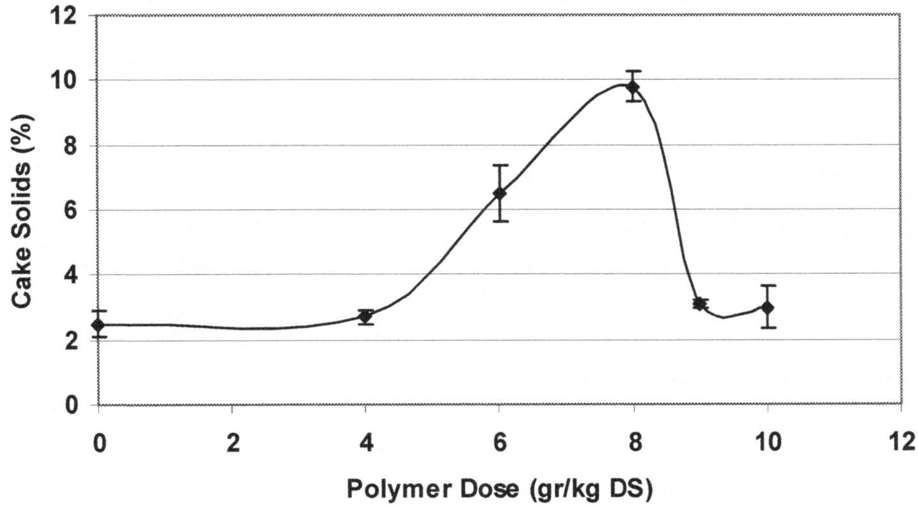

Figure 5-9. Cake Solids Percent of Sludge at Different Polymer Doses.

5.2 Effect of Mixing Intensity and Time on Rheological Measurements and Optimum Polymer Dose

We also conducted experiments to determine the effect of mixing intensity during polymer conditioning of sludge on rheological measurements and optimum polymer dose using anaerobically digested sludge. Seven combinations of mixing time and mixing speed were tested to determine the effect of mixing conditions on rheograms, optimum dose, and dewaterability of sludge. These mixing conditions were: 1, 2, 5, and 10 minutes of mixing at 100 rpm; and 2, 5, and 10 minutes of mixing at 300 rpm. A jar tester with four paddles was used to mix the sludge samples. The volume of the sludge samples was 200 mL. Mixing the sludge at 100 rpm for 1 minute was not adequate to thoroughly mix the polymer in the sample; therefore the results from this set are not included in this report. Again all measurements (rheograms, CST, TTF, solids percent of the cake) were performed in triplicate, and the torque, peak torque, and network strength values were calculated as explained previously. A 0.5% Stockhousen 655K polymer solution was used to condition the sludge.

Figures 5-10, 5-11, and 5-12 illustrate the rheograms obtained after conditioning the sludge at 100 rpm for 2, 5, and 10 minutes, respectively. All rheograms show the average of three measurements. (Error bars clutter the figures, and thus were not included.) However, as previously shown, the reproducibility of the measurements was very good and error bars are relatively small.

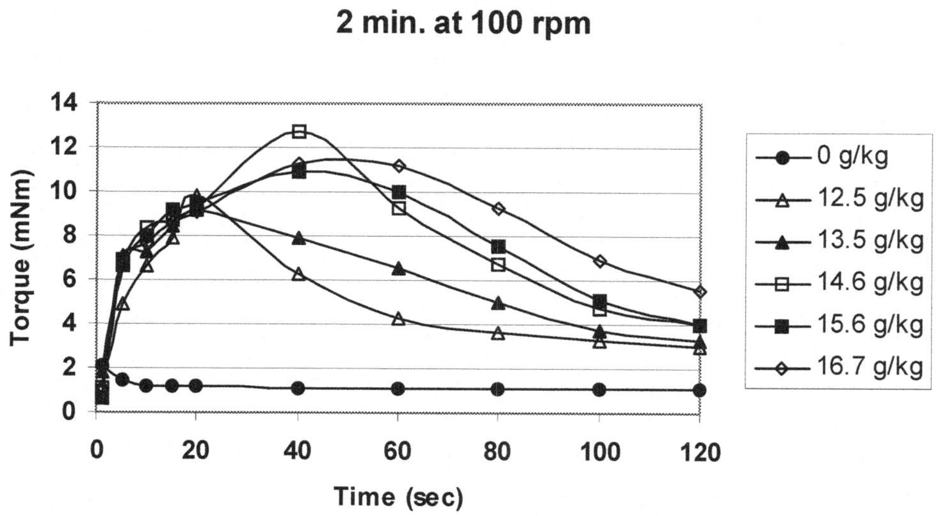

Figure 5-10. Rheograms from Anaerobically Digested Sludge Conditioned at Different Polymer Doses at a Mixing Intensity of 2 Minutes at 100 rpm.

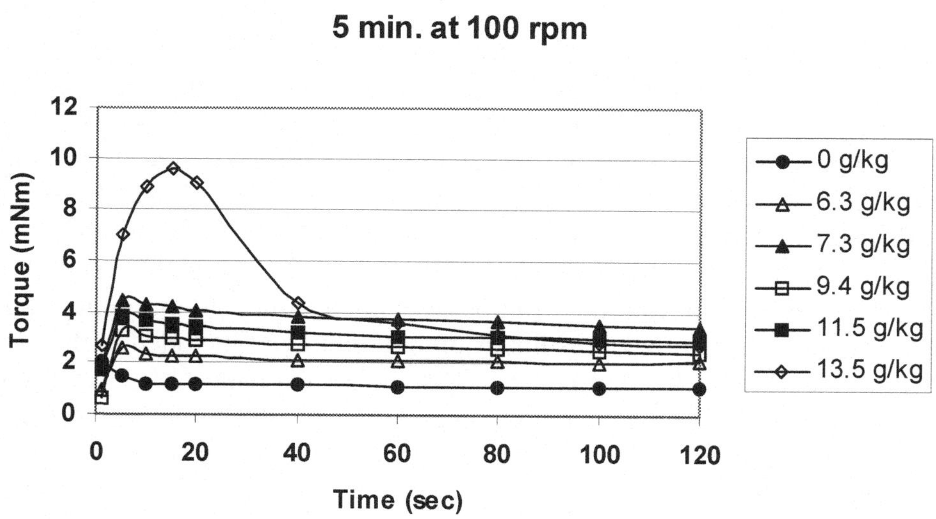

Figure 5-11. Rheograms from Anaerobically Digested Sludge Conditioned at Different Polymer Doses at a Mixing Intensity of 5 Minutes at 100 rpm.

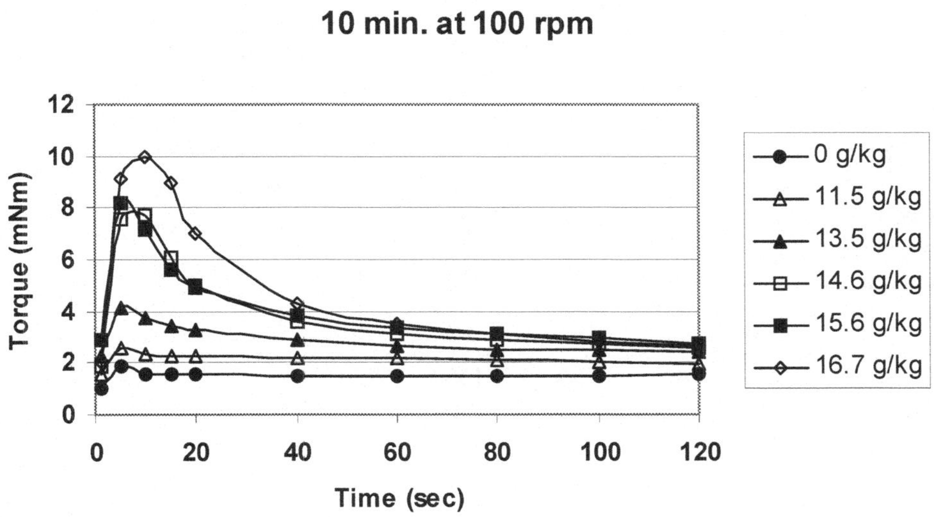

Figure 5-12. Rheograms from Anaerobically Digested Sludge Conditioned at Different Polymer Doses at a Mixing Intensity of 10 Minutes at 100 rpm.

A mixing of 2 minutes at 100 rpm was not strong enough to blend the polymer thoroughly in the sample (Figure 5-10), and the undissolved polymer content caused formation of large erratic peaks. These peaks disappeared at 5- and 10-minute mixing times (Figures 5-11 and 5-12). Longer operation times (impeller mixing time) resulted in decreased torque readings at all of the mixing combinations. For example, the measured torque values at 40 seconds were significantly higher than the measured torque values at 120 seconds. In addition, the tails of the rheograms measured at different doses started to converge at prolonged operation times (such as 120 seconds) decreasing the difference between the torque values at different polymer doses. This is because extended mixing introduced during the operation time is likely to break up the flocs, and expose more particles to the polymer and increase the polymer demand for conditioning. The effect of the polymer on the network strength will not be as obvious as before, and the difference between the rheograms will start to disappear. Thus, for a particular mixing (Gt), the operation time should be neither too short nor too long to calculate a meaningful network strength.

Figures 5-13, 5-14, and 5-15 show the rheograms at 300 rpm for 2, 5, and 10 minutes, respectively. The conditions used to mix the polymer also seem to affect the shape of the rheograms and torque values measured by the rheometer. This can be seen by comparing the rheograms from 100 rpm at 2, 5, and 10 minutes (Figures 5-10 to 5-12), to the rheograms from 300 rpm at 2, 5, and 10 minutes (Figures 5-13 to 5-15). At increased mixing times, the torque values start to decrease at the same impeller speed, and the rheograms start to converge. This can be explained by the increase in the polymer demand to achieve the same level of conditioning with increasing mixing times.

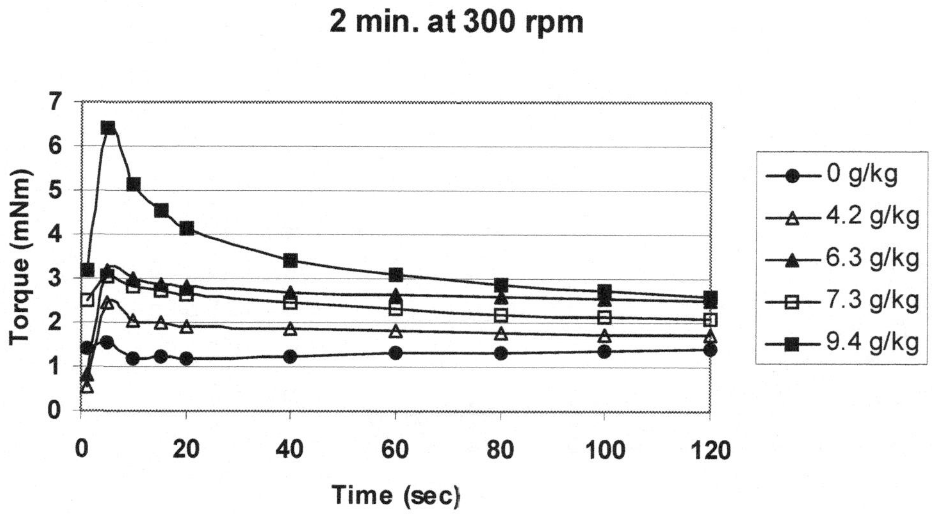

Figure 5-13. Rheograms from Anaerobically Digested Sludge Conditioned at Different Polymer Doses at a Mixing Intensity of 2 Minutes at 300 rpm.

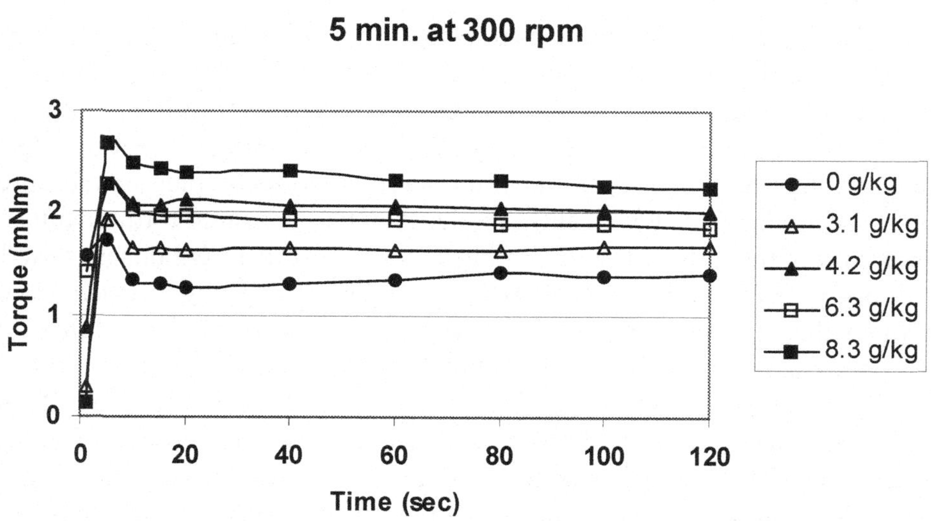

Figure 5-14. Rheograms from Anaerobically Digested Sludge Conditioned at Different Polymer Doses at a Mixing Intensity of 5 Minutes at 300 rpm.

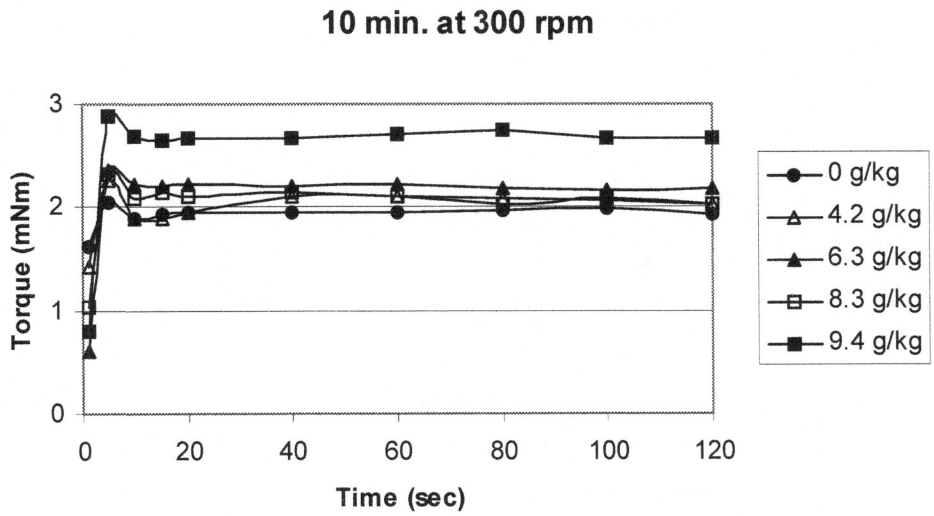

Figure 5-15. Rheograms from Anaerobically Digested Sludge Conditioned at Different Polymer Doses at a Mixing Intensity of 10 Minutes at 300 rpm.

The results of the dewaterability tests (CST, TTF, percent cake solids) from the 100 rpm and 300 rpm sets show that the optimum mixing time is 5 minutes at 100 rpm, and 2 minutes at 300 rpm. Figures 5-16 and 5-17 show the results from the CST tests (results from filtration and percent cake solids are not shown, but they are similar to the CST results). Mixing sludge at 100 rpm for 5 minutes resulted in the lowest polymer consumption to achieve the same level of dewaterability compared to 2 minutes and 10 minutes of mixing times. At 300 rpm, on the other hand, 2 minutes of mixing resulted in the lowest polymer use. Higher mixing times increased the required polymer dose to achieve the same level of dewaterability. These results show that at any mixing speed, there is an optimum mixing time. Under or over mixing at a certain speed increases the optimum dose to condition the sludge.

The results from the dewaterability tests also show that the optimum polymer dose at 100 rpm (around 15 g/kg DS) is much higher than the optimum polymer dose at 300 rpm (around 8 g/kg DS). The required polymer dose for optimum dewatering is different at each mixing speed and time combination, and it is a challenge to determine the combination that would result in the lowest polymer consumption. Comparison of the optimum doses from 100 rpm and 300 rpm sets showed that the lowest polymer dose (8 g/kg DS) was achieved at 300 rpm-2 minutes combination. Thus, we concluded that this represented the optimum mixing conditions for the particular sludge and polymer used in this study. However, the optimum Gt will be different for different polymers and sludge. These results suggest that excessive mixing should be avoided during sludge conditioning in treatment plants as it is likely to increase the polymer dose required for optimum conditioning.

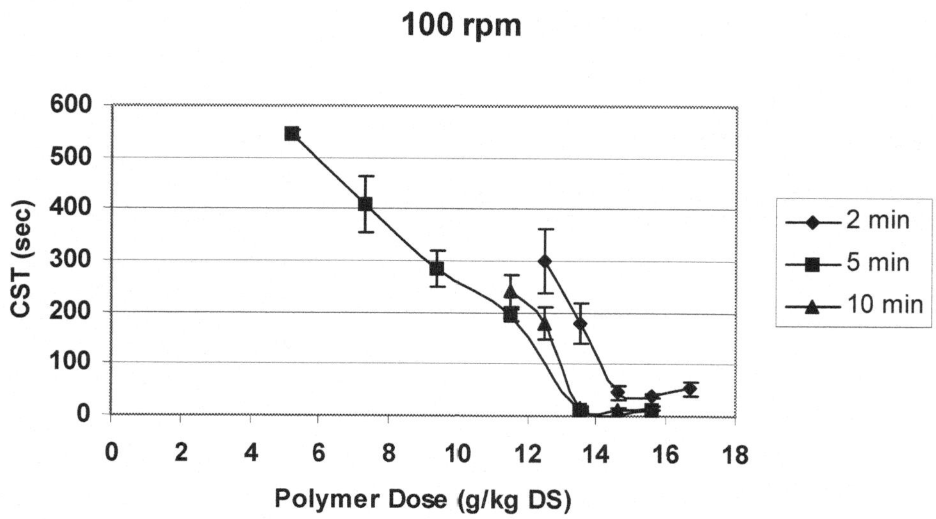

Figure 5-16. CST of Anaerobically Digested Sludge Samples Mixed at Different Shear Times at 100 rpm.

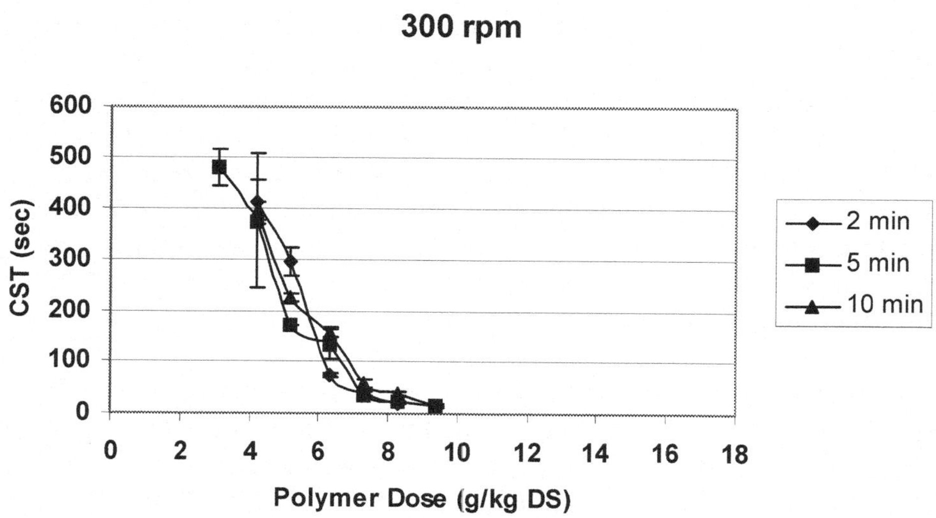

Figure 5-17. CST of Anaerobically Digested Sludge Samples Mixed at Different Shear Times at 300 rpm.

The network strengths calculated by using the area under the rheograms are illustrated in Figures 5-18 and 5-19, where the dip in the network strength at around the optimum polymer dose at each mixing speed and time is evident. The network strength decreased at extended mixing times, and the magnitude of the dip observed at the optimum polymer dose decreased as well. Thus, it is important to work at the optimum mixing range when determining the optimum polymer dose using rheological characteristics of sludge. If the mixing conditions are too high, it might be very difficult to observe the dip, as in the case of 10 minutes of mixing at 300 rpm.

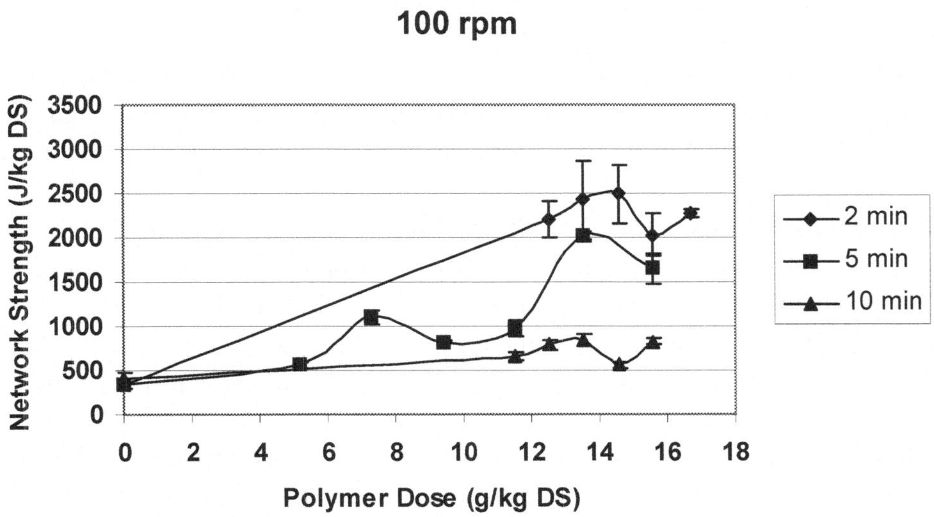

Figure 5-18. Network Strength Values Calculated for Anaerobically Digested Sludge Samples Mixed at Different Shear Times at 100 rpm.

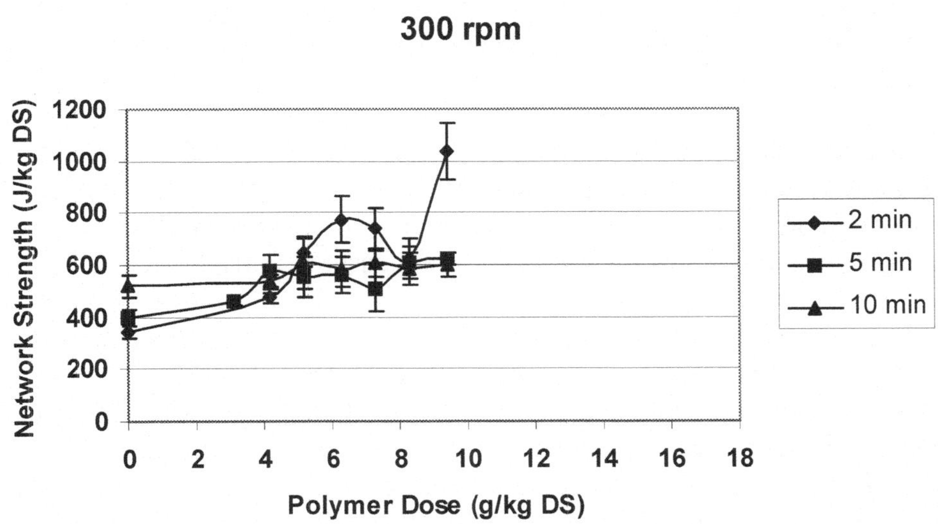

Figure 5-19. Network Strength Values Calculated for Anaerobically Digested Sludge Samples Mixed at Different Shear Times at 300 rpm.

5.3 Effect of Polymer Type on Rheological Measurements and Optimum Polymer Dose

The objectives of these experiments were to evaluate the effect of different polymer types on rheological measurements and the optimum polymer dose.

Five polymers (C-1598, SD-2080, SD-2081, SD-2083, SD-2085) with different degrees of polymer structure were used in the experiments. All of the polymers have a molecular weight of 5,000,000. The specifications of these polymers are listed below.

- Superfloc SD-2080 Flocculant: Cationic polyacrylomide in water-in-oil emulsion
- Superfloc SD-2081 Flocculant: Cationic polyacrylomide in water-in-oil emulsion
- Superfloc SD-2083 Flocculant: Liquid cationic polyacrylomide
- Superfloc SD-2085 Flocculant: Liquid cationic polyacrylomide
- Superfloc C-1598 Flocculant: Cationic polyacrylomide in water-in-oil emulsion

The manufacturer (Cytec Industries Inc., Stanford, Connecticut) noted that the differences in the degree of polymer structure correlate with the differences in the final floc strength for certain types of sludge. All of these polymers have a nominal cationic charge of 55 mol% and a similar range of molecular weight.

The degree of polymer structure increases in the following order: C-1598 < SD-2080 < SD-2081 < SD-2083 < SD-2085. The polymer solutions were prepared at strength of 1% using the following method according to the instructions of the manufacturer: 297 mL of water was placed in a blender. The blender was switched on and 3 mL polymer emulsion was added to the water. The mixture was blended for 20 to 30 seconds.

Rheological measurements were made at 300 rpm impeller speed using anaerobically digested sludge. The polymer was mixed using a jar-tester at 300 rpm for 2 minutes. The optimum dose was determined for each polymer, and rheological characteristics and dewaterability of the samples were measured in triplicate.

Due to the large number of rheograms produced, the raw rheograms obtained from each polymer at different doses are not shown here. Instead, the torque values measured at 40 seconds for each dose and polymer are illustrated in Figure 5-20, which shows the changes in the rheological parameters with increasing polymer dose. The accompanying CST, TTF, and cake solids data for each polymer are illustrated in Figures 5-21, 5-22, and 5-23, respectively. The network strengths of the sludge samples are illustrated in Figures 5-24 through 5-28.

The required polymer dose to achieve an acceptable dewaterability was different for each polymer, but the decrease in the network strength seemed to correlate well with the optimum dose. The dewaterability data indicate that the optimum dose was around 20 g/kg DS for the polymer SD-2080 (Figures 5-21 to 5-23). The decrease in the network strength was also observed at 20 g/kg DS (Figure 5-24). Polymer SD-2081 produced two dips on the rheograms. The first dip was at 25.8 g/kg DS, and the second dip was at 32.8 g/kg DS. The first dip corresponded to the polymer dose where a rapid drop in the CST was observed (from 210 to 80 seconds), the second dip corresponded to the real optimum dose where the CST was 17 seconds, and the filtrate volume and the cake solids were highest of all doses (Figures 5-21 to 5-23). The improvement in the dewaterability of sludge was gradual with polymer SD-2083, and the higher polymer doses resulted in better dewaterability (Figures 5-21 and 5-22). The dip in the network strength was observed at 28.1 g/kg DS (Figure 5-26), and CST of the sludge was down to 20 seconds at this dose. However, the cake solids percent did not show a significant improvement with this polymer. When polymer SD-2085 was used to condition the sludge, the decrease in the network strength started at 39.8 g/kg DS (Figure 5-27), and this dose resulted in good dewaterability (Figures 5-21 to 5-23). The polymer demand was lowest when polymer C-1598 was used. The dip in the network strength was observed at 11.7 g/kg DS and CST was 68.6 seconds at this dose (Figure 5-28 and Figure 5-21). At the next higher dose tested, 14.1 g/kg DS, the CST dropped down to 21.4 seconds, and this seemed to be the optimum dose.

To compare the effectiveness of the polymers, we compared the amount of polymer needed to achieve the same level of dewaterability by comparing the data from the CST, filtration, and cake solids percent tests. The polymer consumption increased in the following order: C-1598 < SD-2080 < SD-2081 < SD-2083 < SD-2085. According to the manufacturer's information the increase in the degree of polymer structure followed the same order.

Overall, for all of the polymers tested, the decrease observed in the network strength correlated well with the optimum dose which was determined from the results of the CST, filtration, and percent cake solids tests. Two dips were observed with one of the polymers, the first dip indicating the critical point where rapid improvement in the dewaterability was observed, and the second dip indicating the optimum dose.

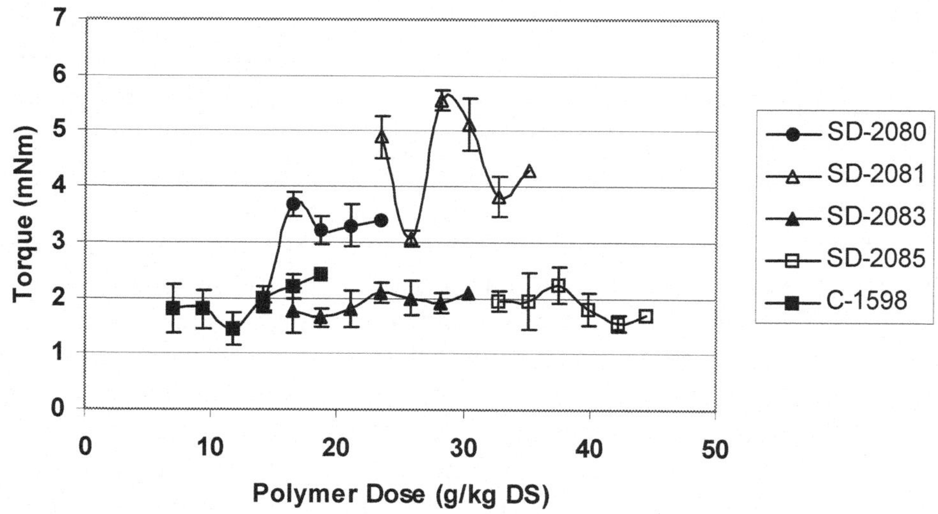

Figure 5-20. Torque Values for Anaerobically Digested Sludge Conditioned with Five Polymers at Different Doses.

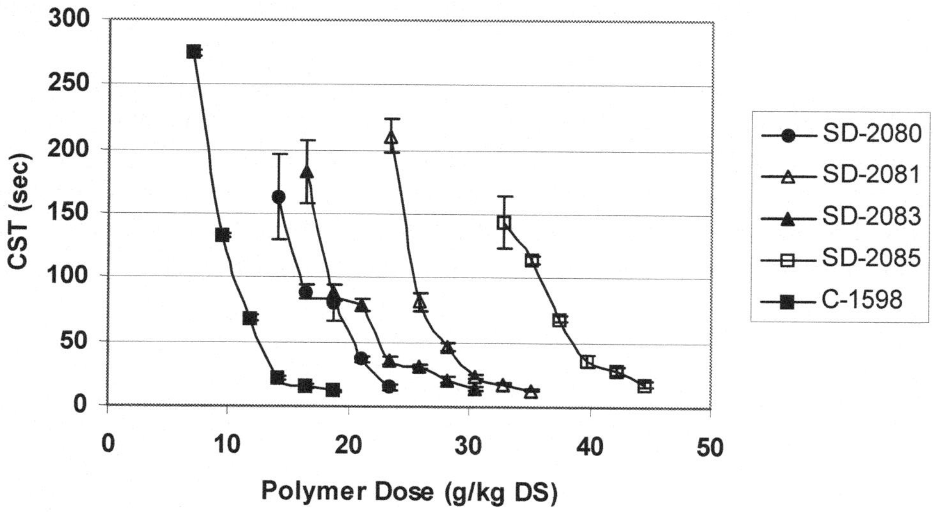

Figure 5-21. CST of Anaerobically Digested Sludge Conditioned with Five Polymers at Different Doses.

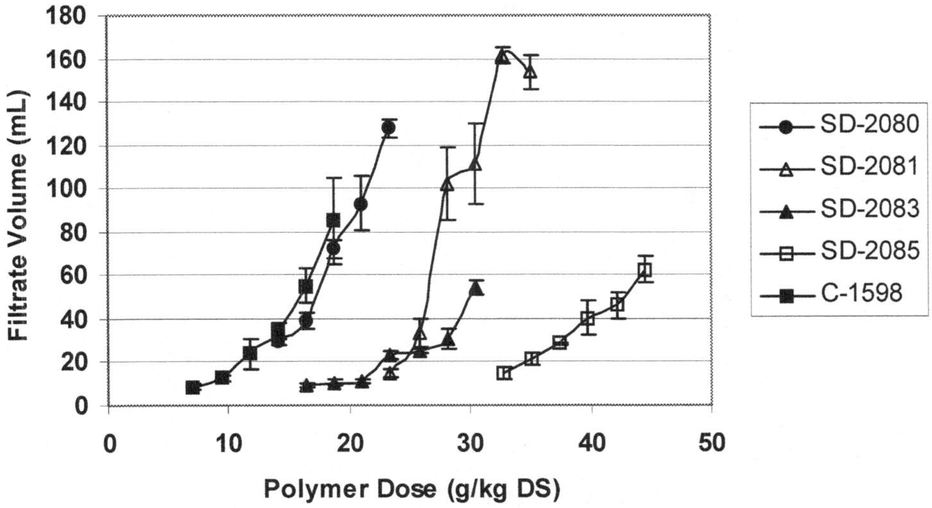

Figure 5-22. Filtrate Volume of Anaerobically Digested Sludge Conditioned with Five Polymers at Different Doses.

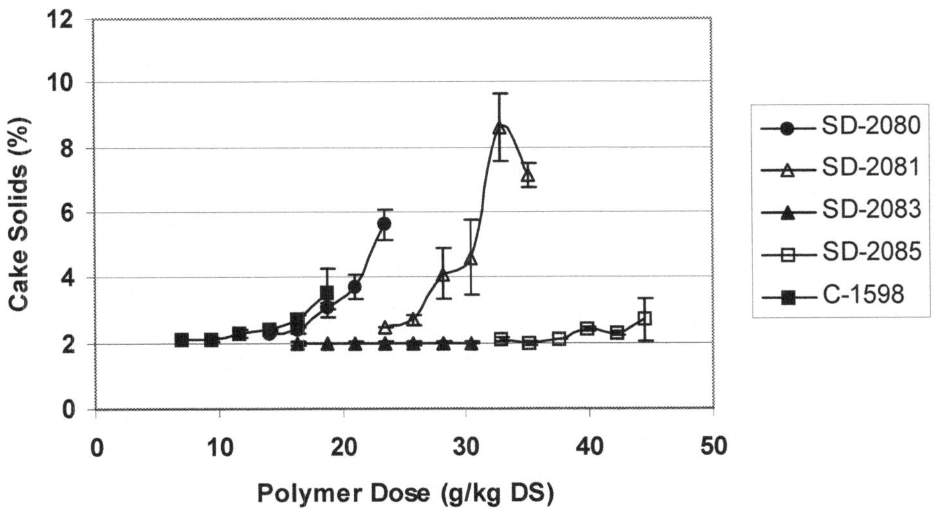

Figure 5-23. Cake Solids Percent of Anaerobically Digested Sludge Conditioned with Five Polymers at Different Doses.

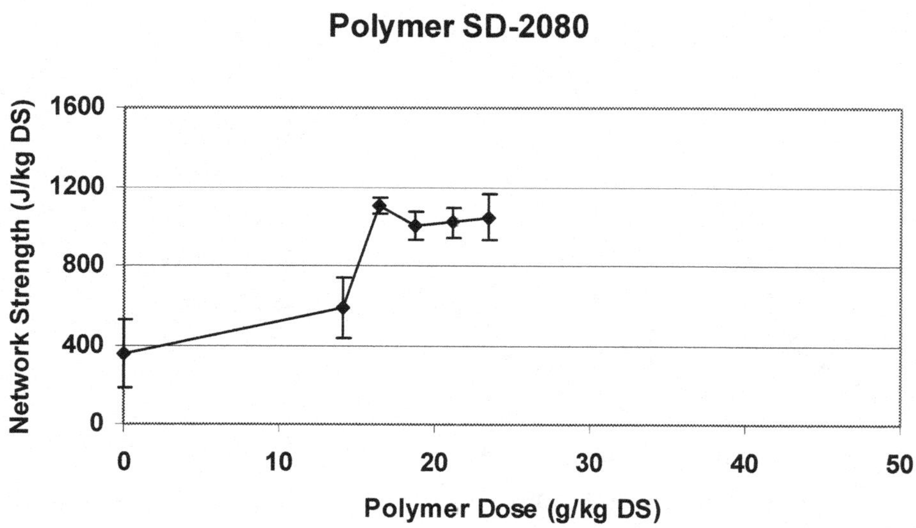

Figure 5-24. Network Strength of Anaerobically Digested Sludge Conditioned with Polymer SD-2080. Optimum Polymer Dose is ~20 g/kg DS.

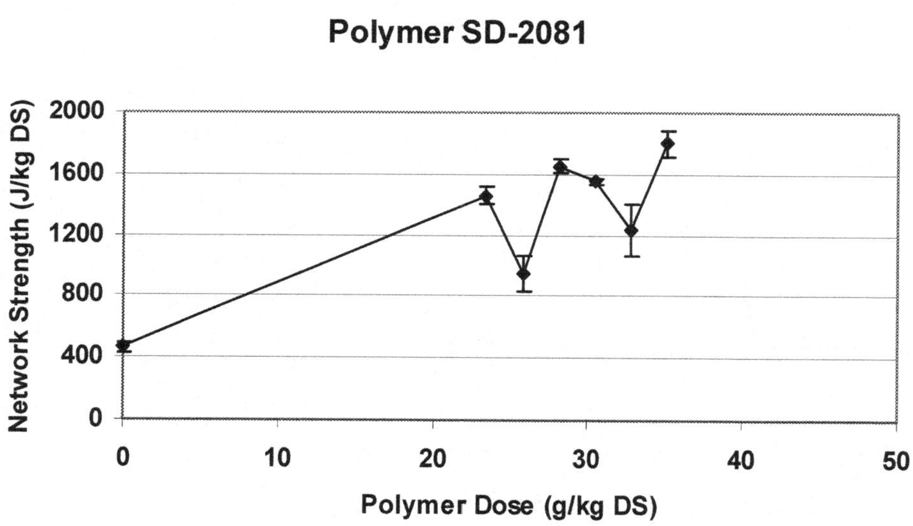

Figure 5-25. Network Strength of Anaerobically Digested Sludge Conditioned with Polymer SD-2081. Optimum Polymer Dose is ~32.8 g/kg DS.

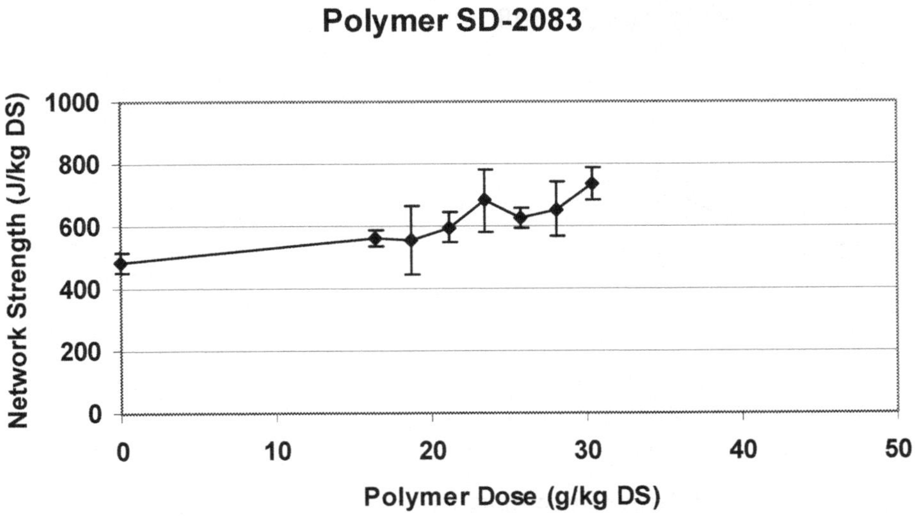

Figure 5-26. Network Strength of Anaerobically Digested Sludge Conditioned with Polymer SD-2083. Optimum Polymer Dose is ~28.1 g/kg DS.

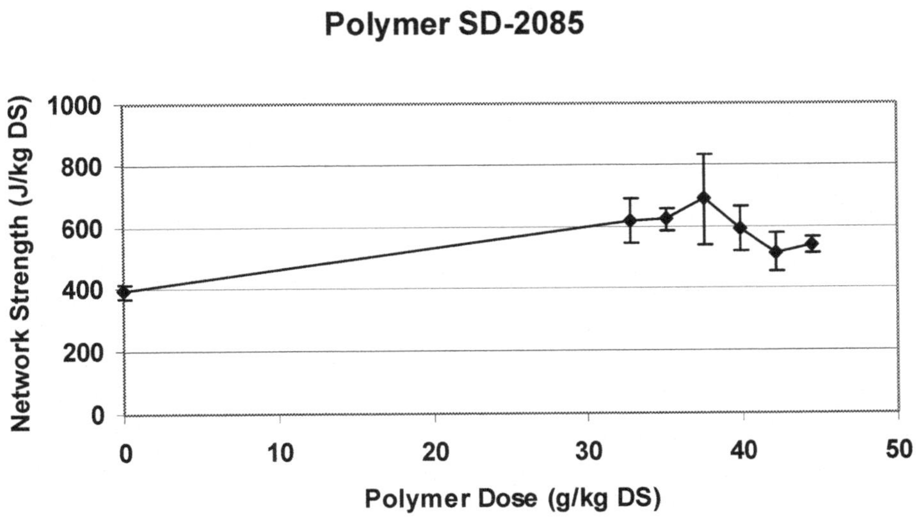

Figure 5-27. Network Strength of Anaerobically Digested Sludge Conditioned with Polymer SD-2085. Optimum Polymer Dose is ~39.8 g/kg DS.

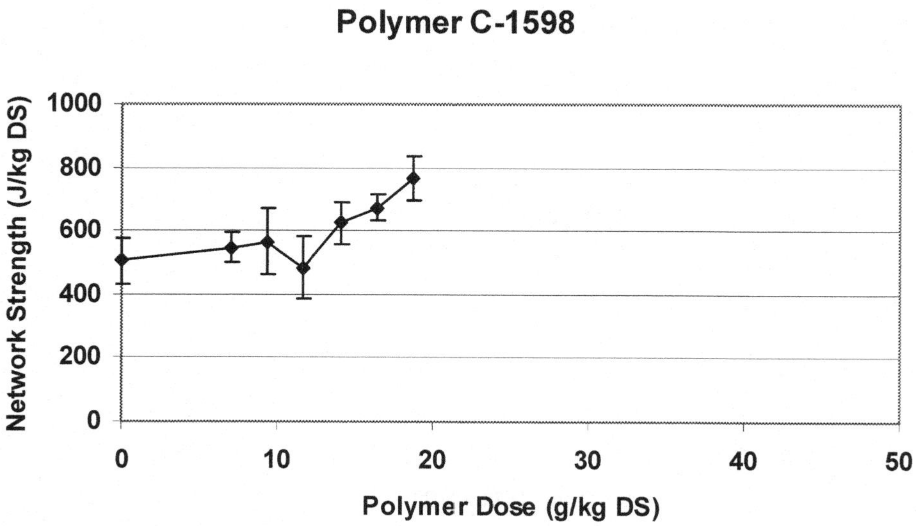

Figure 5-28. Network Strength of Anaerobically Digested Sludge Conditioned with Polymer C-1598. Optimum Polymer Dose is ~14.1 g/kg DS.

CHAPTER 6.0

FULL-SCALE TESTS

6.1 Overview

The results from the laboratory tests showed that the method we proposed to measure the network strength and determine the optimum dose using rheological characteristics of sludge works well. The objective of the fieldwork was to verify and validate the principle that the "dip" observed in the network strength as a function of the polymer dose indicates the optimum dosage. The TS in the dewatered sludge and the liquid stream was used as an indication of the dewatering/thickening performance. The CST of the conditioned sludge was also used as a parameter of optimum conditioning. The fieldwork started early in October of 2002 following equipment malfunction due to transporting the rheometer from Duke University in North Carolina to North American Technology Center (NATC) in New Jersey. We are grateful to U.S. Koei for promptly providing a back up unit that allowed completion of this important part in a timely fashion.

Full-scale testing was conducted at the 100 MGD Wilmington Pollution Control Facility operated by Veolia Water North America Operating Services. Testing was conducted on both centrifuge and gravity belt thickeners. The dose-response testing protocol as documented by Dentel et al. (1995) was used in these experiments. Generally, the procedures force incremental changes in the polymer dose, allow the thickening or dewatering device to stabilize with the polymer feed, and sample the output for appropriate measurements. All tests including CST, rheology, and TS, were conducted on site to eliminate any changes that might alter the sludge characteristics during transport to a different site (Abu-Orf et al., 2002). TS and CST were conducted in triplicates, and rheological measurements were conducted four times.

6.2 GBT Tests

GBTs were used at the Wilmington facility to thicken a blend of waste activated sludge (WAS) and primary sludge prior to feeding the anaerobic digestion system. The facility has two, 3-meter Roediger GBTs (Roediger Model 832.3, 3.2-m width) where one usually operates continuously and the second is used as a backup. The facility uses a Mannich polymer (PolyDyne C-321) at a 0.5% concentration for conditioning the sludge.

6.2.1 Procedure

The polymer feed rate was reduced to the lowest value that allowed the flocculated sludge to stay on the belt without overflowing from the sides of the GBT. The GBT was allowed to stabilize with the polymer feed rate (~10 minutes), and samples of the filtrate and thickened sludge were collected for total solids determination, which was conducted using the plant's facilities. Conditioned sludge was collected for CST and rheology measurements right from the outlet to the drainage zone using a typical sampler cup. Four replicates of the rheology were measured at each polymer dose. Both measurements were conducted instantaneously on site. The polymer feed was increased incrementally to the highest rate possible, the GBT was allowed to stabilize with each polymer feed rate, and the same samples were collected for analysis.

6.2.2 Results

The results of the GBT dose response testing are shown in Figures 6-1 through 6-3. Figure 6-1 shows the CST and the filtrate solids (TS) results. All the average CST results were less than 8 seconds, which indicates that the sludge exhibited good dewaterability. If we consider the lowest CST values to represent the best conditions for sludge thickening, a dose of 20 to 22 g/kg DS is considered the optimum dose range. The same dose range could be considered the optimum range from the filtrate percent TS graph in Figure 6-1.

Figure 6-2 presents the thickened sludge TS and the network strength at different times during shear application. At the 20 to 22 g/kg DS dose range, the thickened solids were at the highest values of ~7.5%, which confirms the consideration of this dose range as an optimum. Figures 6-3 provides another representation of the network strength of the conditioned sludge after 40 seconds of shear application, which was calculated as previously stated in Chapter 2.0 using the area under the torque-time relationship. The network strength increased with increasing the polymer dose and decreased at the polymer dose range between 19 and 21 g/kg DS, which was determined to be the optimum dose range from the CST, filtrate, and thickened sludge TS. At higher polymer dosages, the network strength started to increase again. Accordingly, the network strength of the conditioned sludge could be used to identify the optimum polymer dose during thickening, where the optimum dose is indicated by a "dip" in the network strength polymer relationship.

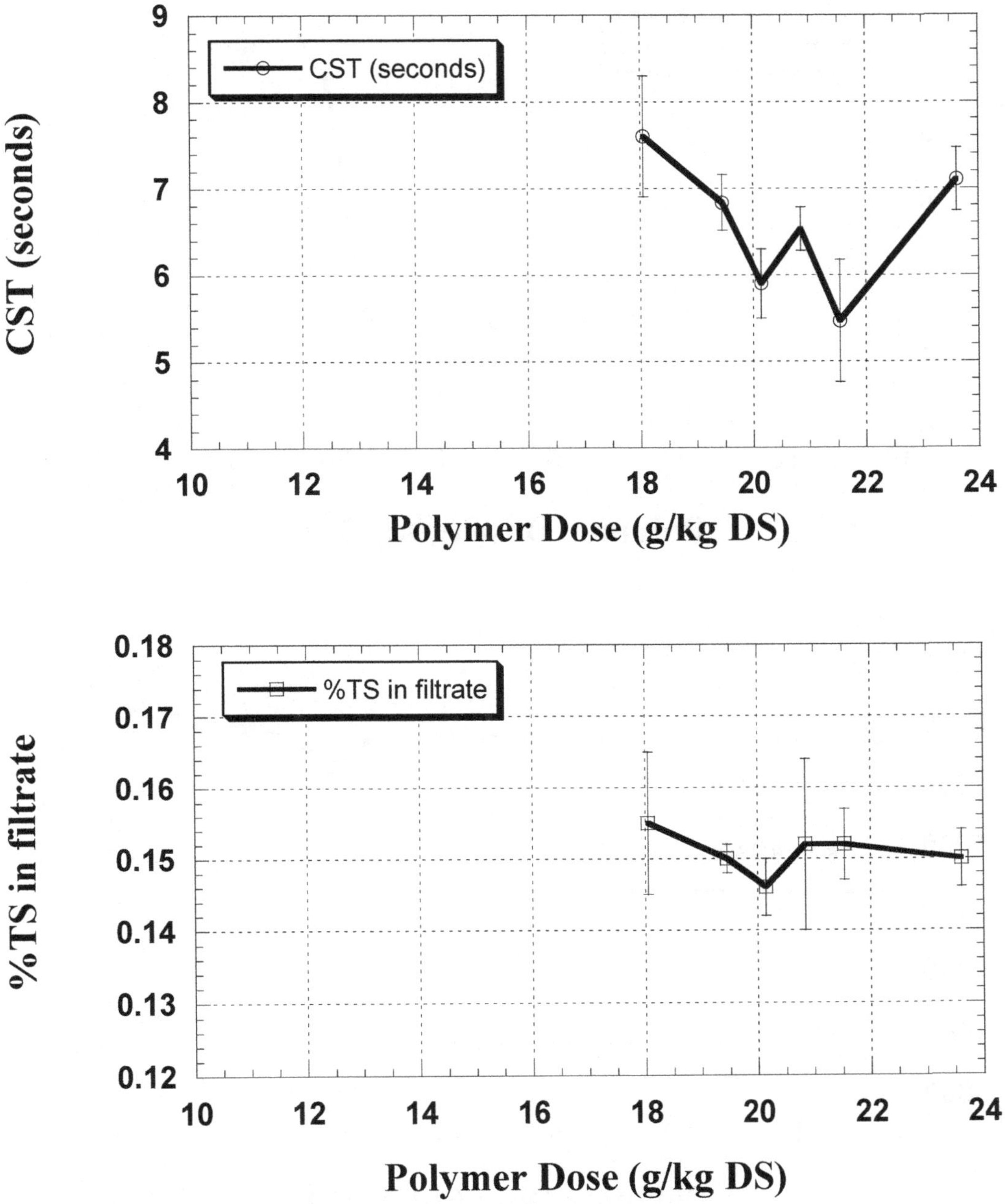

Figure 6-1. CST of Conditioned Sludge and Percent Total Solids in Filtrate as a Function of Polymer Dose.

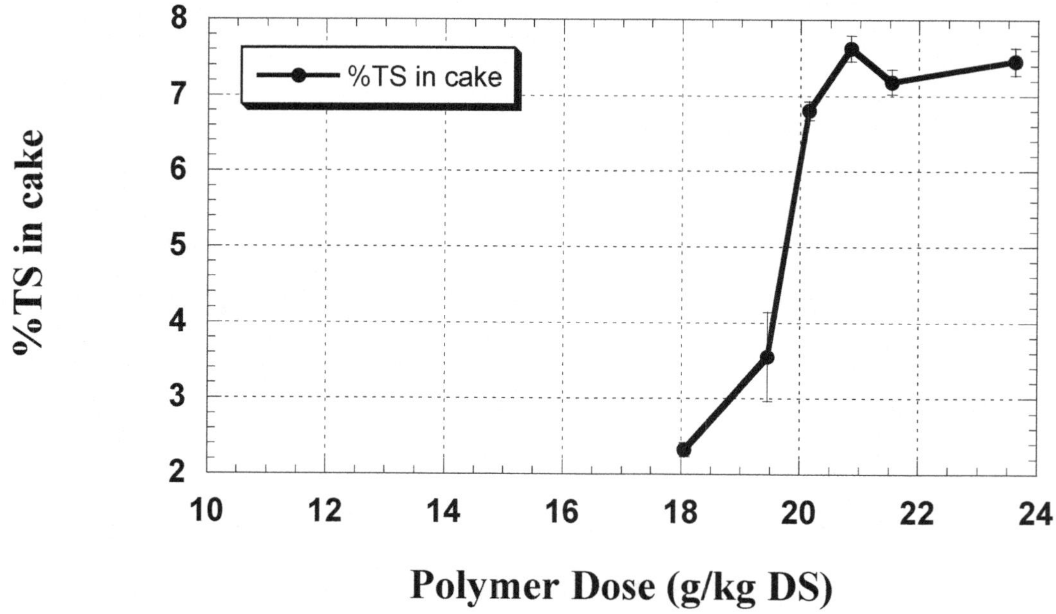

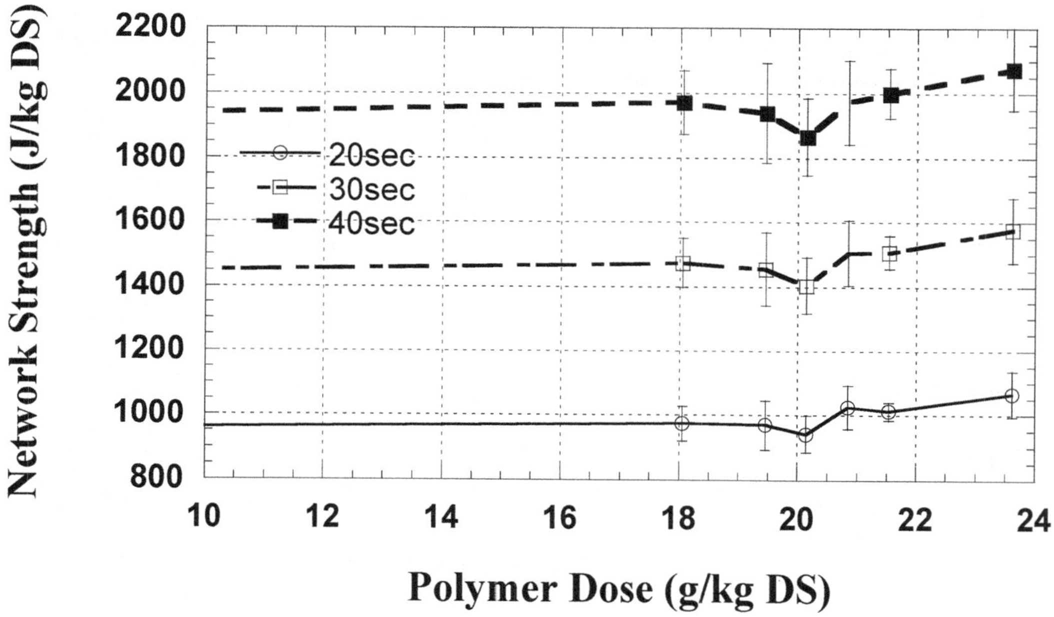

Figure 6-2. Percent Total Solids in Thickened Sludge and Network Strength as a Function of Polymer Dose.

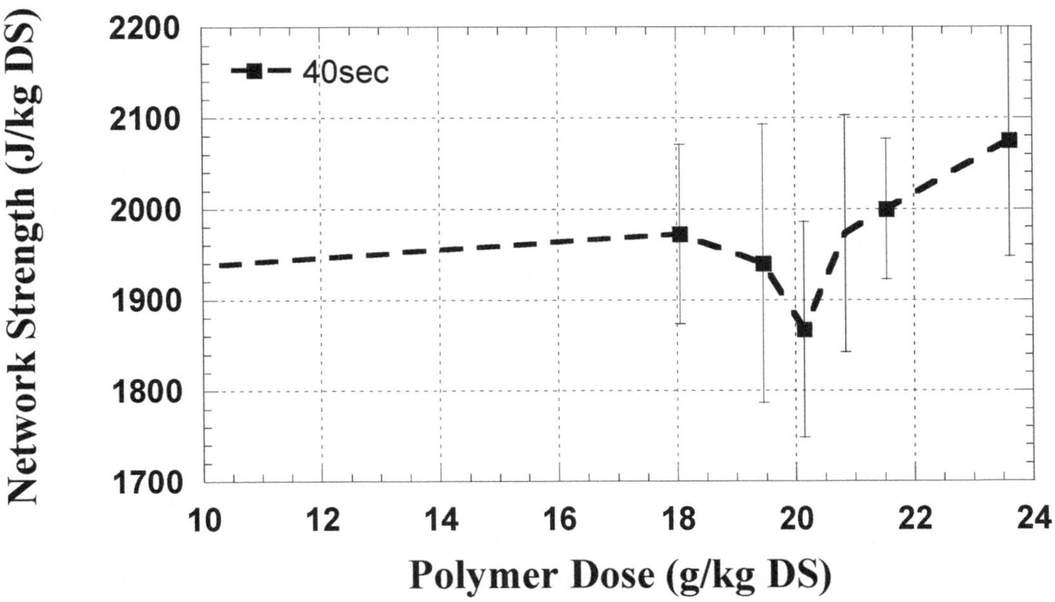

Figure 6-3. Network Strength at 40 Seconds of Shear Application as a Function of Polymer Dose.

6.3 Centrifuge Performance Tests

At the Wilmington facility, a centrifuge is used to dewater anaerobically digested sludge prior to further alkaline stabilization and disposal. The facility has one operational centrifuge (Bird Model 6150, designed for a maximum feed rate of 400 gpm), which usually operates continuously. The facility uses Polydyne MA 105 dry polymer at a 0.5% concentration to condition sludge solids for dewatering.

6.3.1 Procedures

The polymer feed rate to the centrifuge was increased to about 120% of its value as it was operated. The centrifuge was allowed to stabilize with the new polymer feed (~30 minutes), and samples of the filtrate and cake were collected for TS determination, which was conducted on site. Conditioned sludge was collected from the sampling port for CST and rheology measurements just before the sludge entered the centrifuge. Both measurements were conducted simultaneously on site. The polymer feed was decreased incrementally to its lowest rate possible without adversely affecting the operation of the centrifuge, which was determined by the dark color of the centrate as indicated by the increased solids in the centrate. The centrifuge was allowed to stabilize with each polymer dose, and samples were collected for analysis.

6.3.2 Results

Figures 6-4 and 6-5 present the results from the full-scale centrifuge tests. Figure 6-4 shows the CST and centrate TS and Figure 6-5 shows the cake solids and the conditioned sludge network strength. The CST data is inconclusive and do not indicate the optimum dose range for optimum centrifugation. Previously, Abu-Orf and Dentel (1997a) observed similar trends where the CST failed to indicate the optimum dose range from full-scale conditions. The TS of the centrate kept decreasing with increasing the polymer dose, and the cake TS kept increasing with increasing the polymer dose. Consequently, the solids results cannot be used to precisely determine the optimum polymer dose for the centrifugation.

A look at the average network strength of conditioned sludge shows the same trend as previously described in this research, where a "dip" is obvious at a polymer dose range of 19 to 21 g/kg DS. We cannot determine whether this is the optimum dose from the solids results of the centrifuge. However, previous results indicate it may be. We have recently conducted experiments with the Wilmington centrifuge and concluded that the centrifuge is not operating properly. In fact, the plant is in the process of purchasing two new centrifuges, and pilot testing was to be conducted in November 2002. Accordingly, results from this centrifuge are not reliable.

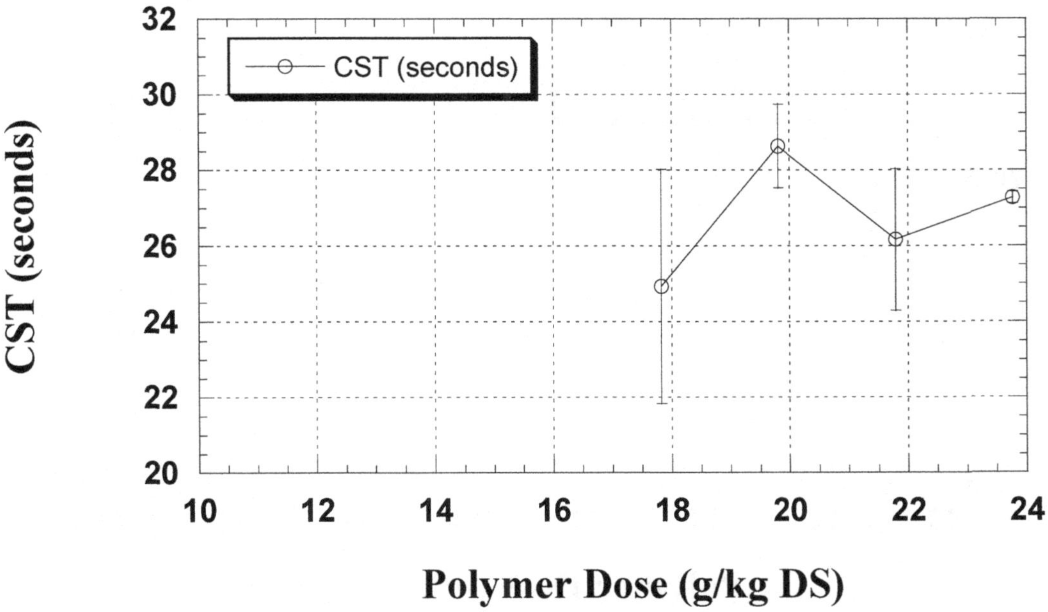

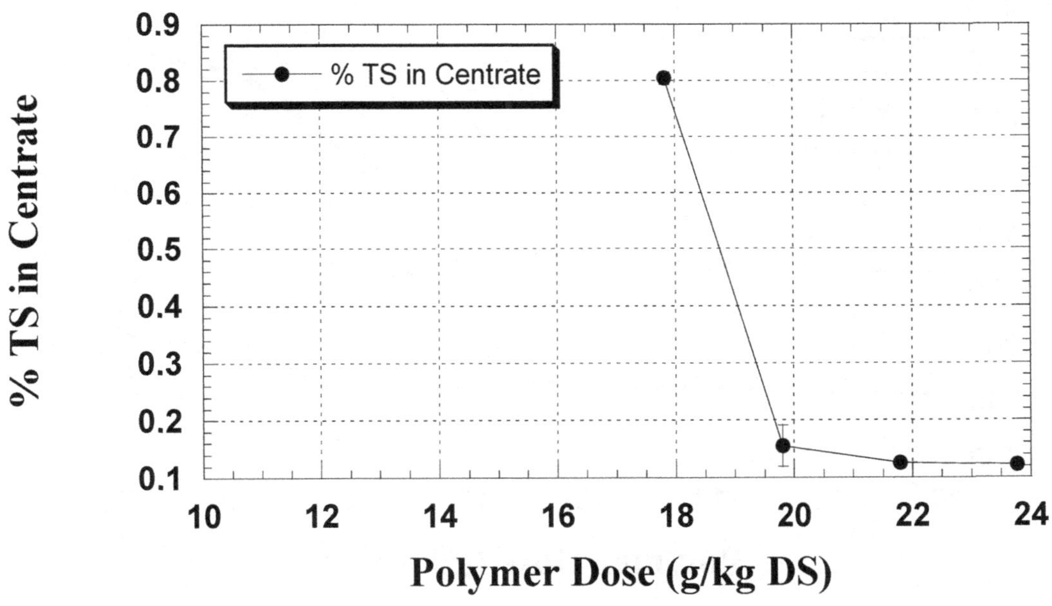

Figure 6-4. CST of Conditioned Sludge and Percent Total Solids in Filtrate as a Function of Polymer Dose.

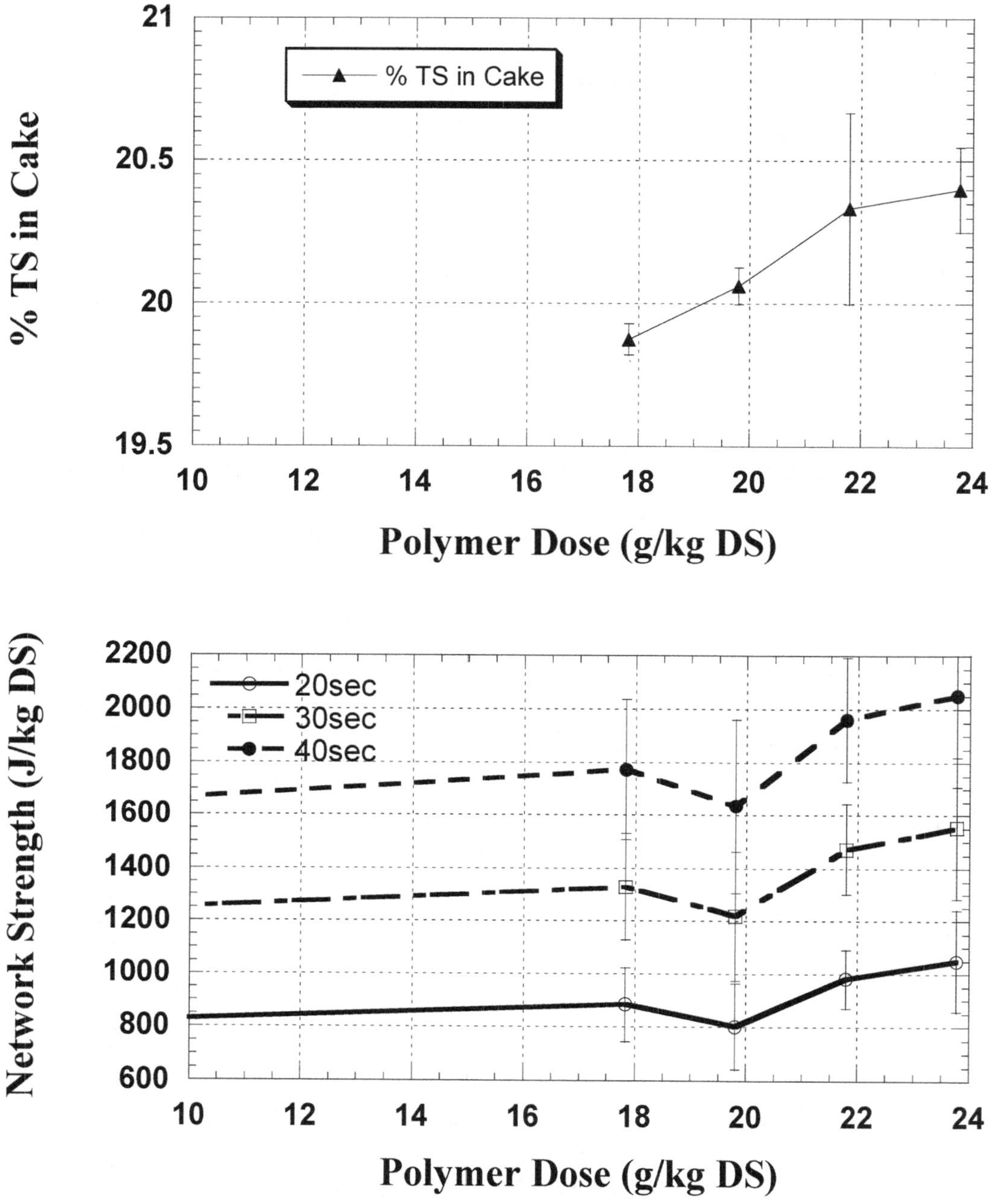

Figure 6-5. Percent Total Solids in Cake and Network Strength as a Function of Polymer Dose.

CHAPTER 7.0

DISCUSSION

7.1 Optimum Impeller Speed

The results presented in Chapter 4.0 indicated that the optimum impeller speed was observed to be the lowest speed that produced a significant signal output. Higher speeds decreased the sensitivity of the measurements due to the increased turbulence in the beaker and increased the magnitude of the error bars. The beaker effect should also be considered in choosing the optimum speed. When impeller speed is high, sludge flocs tend to move away from the impeller toward the beaker walls. This reduces the ability of the Floccky rheometer to break-up the flocs that get in to contact with the impeller and therefore lower the accuracy of the measurements. Remember, the single most important factor that determines the optimum speed is the solids content of the sludge. Optimum speed increases with decreasing solids content.

This study used activated, aerobically digested, anaerobically digested, and alum sludge. The optimum impeller speeds for the sludge types used in this study were determined as follows: 300 rpm for aerobically and anaerobically digested sludge, and 350 rpm for activated sludge and alum sludge. The optimum speed for a particular sludge may change, of course, depending on the solids content of the sludge on that day.

The speed of the jar tester should also be considered in choosing the optimum speed. In theory, the impeller speed should not be higher than the jar-tester speed. For example, if the mixing of the polymer was done at 200 rpm, and the rheology was measured at 300 rpm, this meant a more powerful mixing would be applied to the sample during the measurement. The additional mixing introduced to the sample would change the sample characteristics and the optimum dose. Thus, the sample would no longer represent its original characteristics.

7.2 Implication to Dewatering

The desired or required dewaterability from a dewatering device is the overall output of the process. The factors that affect the desired dewaterability are almost always specific to the dewatering device's ability to squeeze or withdraw water from the sludge without applying too much energy to cause floc deterioration. Current practices utilize adjustments to the device's operation on a semi-continuous, after-the-fact basis through operator intervention. The factors that influence the properties of the flocculated sludge to be subjected to dewatering forces (e.g., how the sludge is flocculated in the first place) are not usually optimized for the dewatering process, but instead for the thickening process. Plant operators should be able to design the suspension's strength to *match* the shear strength that the suspension's flocs will be subjected to during the dewatering process in order to achieve the desired dewaterability. The use of rheology to measure the suspension's strength is an appropriate tool.

The factors that affect rheology measurements were examined and once understood they were used in a proactive manner. These factors include: the suitability of the rheology device, and the rate of applying the shear for rheology measurements (impeller speed). The factors that affect the network strength include the conditioner type, the mixing conditions during conditioning, the inherent properties of the sludge (e.g., cations content, biopolymer content, TS), as well as other properties. This research showed that changing the mixing conditions during conditioning and using different polymers for conditioning will affect the network strength. Accordingly, plant operators can alter these factors to produce stronger or weaker networks depending on the required dewatering output.

The results presented in this report offer a protocol for measuring comparative sludge network strength as indicated by the total energy dissipated in the sludge system while applying shear forces in rheological measurements. One important finding of this research is that the energy within the network correlated well with the filtration and to some extent the CST results for indicating the optimum dose in laboratory experiments. The optimum dose is indicated by a "drop" in the energy with increasing the polymer dose at the optimum level. The decrease observed in the energy of the network at the optimum dose can be explained as follows. Water associated with sludge flocs can be categorized in three different groups (Vesilind, 1994; Kopp and Dichtl, 2001): free water, interstitial water, and bound water. Free water is not bound to the particles, interstitial water is bound by capillary forces between flocs, and bound water is tightly held to the surfaces. Consequently, the intensity of the binding energies of these waters is different, highest for the bound water and lowest for the free water. Kopp and Dichtl (2001) reported that the binding energy for free water is less than 0.28 kJ/kg water, and for bound water is higher than 5 kJ/kg water. When the optimum polymer dose is reached during conditioning, some of the water incorporated in sludge flocs (mostly interstitial water) is converted into free water. Since free water has a lower energy than other types of water, the overall network energy will also be lower at the optimum dose. Thus, the optimum dose can be determined by monitoring the energy of the system.

Another physical explanation for this correlation is that flocs within the network exhibit lower resistance when they are optimally conditioned by a polymer, which allows them to release most of their water by filtration while maintaining their structures' integrity and withstanding the shear. This is an important factor for achieving optimum dewaterability, since the dilemma in designing conditioned sludges has been that it is desirable to have strong flocs during dewatering, yet sludges need to yield water. We have arrived at an innovative concept and technology to overcome this dilemma. Ultimately, treatment plant operators will use the technology prior to sludge dewatering to continuously measure sludge network strength, and adjust the upstream wastewater chemical characteristics (such as the cation composition and content), mixing conditions, and polymer dose to control strength for better dewaterability.

7.3 Implications from Full-scale Testing

The results from the full-scale testing, using two types of sludge along with a dewatering and a thickening device are encouraging. The "dip" in the network strength as a function of the polymer dose in both sludges was obvious and detectable by the laboratory bench rheometer. In the case of the GBT, the dip in the network strength correlates with optimum polymer dose.

Two observations can be made from a closer look at the network strength data from the two conditioned sludge types. First, large error bars are observed with both samples with the conditioned sludge samples showing more variability. Second, the centrifugation sampling showed more variability. The nature of sampling explains the second observations. In the case of the GBT, sampling of the conditioned sludge occurred from a large, upflow reservoir prior to flowing to the gravity belt. The flocs from this sampling were merely uniform with less variability. In the case of the centrifugation, sampling occurred from an 8-in. pipe using a 1-in. sampling faucet, which resulted in different floc sizes in the 1 L sample collected. During the experiments we noted that conditioned anaerobically digested sludge floc sizes were larger than the GBT sludge floc sizes with less uniformity in the floc sizes.

The first observation can be attributed to the 200 mL samples used for rheological measurements that different floc sizes may have been present in each. The phenomenon was previously observed by Abu-Orf and Dentel (1999) during laboratory testing and in Abu-Orf and Dentel (1997b) during full-scale sampling of conditioned sludge from the same facility.

To increase reproducibility and reduce variability from sampling different flocs for rheological measurements, it is recommended that the full-scale torque rheometer be used. This prototype operates in a similar fashion to the lab unit and samples conditioned and unconditioned sludge for rheological measurements. However, the full-scale prototype should sample unconditioned sludge, add polymer, and then collect the rheological data. This will assure similar conditioned flocs for each sample or measurement, which should reduce the variability of the network strength. The mixing conditions during polymer conditioning inside the unit must *closely match* the plant mixing conditions prior to the dewatering process. Dentel and Abu-Orf (1993) proposed a model for estimating the mixing conditions in terms of mixing intensity (rpm) and time for sludge conditioning schemes, which can be used for this purpose. This method of increasing the reproducibility, however, will result in increasing the response time for the automatic control loop.

The results from the full-scale dose response testing, although encouraging, document the need for further research. The full-scale rheometer prototype operates exactly like the lab model and uses the same rotor and gap space, however, certain modifications are needed to overcome the lack of reproducibility observed in the full-scale experiments. The research would first use the full-scale rheometer prototype to verify the results from the lab testing and verify the need for these modifications. Once modifications are complete, if necessary, the prototype will be installed in full-scale to monitor the network strength on line. The project team proposed a continuation phase to verify the use of full-scale prototype for automatic control of conditioning and dewatering to WERF.

CHAPTER 8.0

CONCLUSIONS

Specific conclusions of the research presented in this report are:

- A mathematical model for quantification of the sludge network strength in terms of the energy dissipated to the flocculated networks while applying shear is provided based upon the area under the rheogram (shear stress-shear rate relationship and torque-time relationship).

- The mathematical models do not represent the "absolute" network strength, rather a comparative strength of different aggregates using the same instrument and measuring conditions. Thus, the calculated strength from the measurements is relative.

- The torque rheometer used in this study showed good rheogram reproducibility.

- The concentric cylinder rheometer still faced with reproducibility problems in rheograms even though the whole conditioned sludge sample was used in rheological measurements.

- The optimum impeller speed was observed to be the lowest speed that produced a significant signal output. Higher speeds decreased the sensitivity of the measurements due to the increased turbulence in the beaker. Optimum speed increases with decreasing solids content.

- A correlation between the network strength and the polymer dose was found in this study. The network strength increased as the polymer dose increased, however, at the optimum dose a decrease (dip) in the measured network strength was observed. At high polymer doses, the network strength increased again. The decrease in the network strength correlated well with results from the dewaterability tests (CST, filtrate volume, and percent cake solids).

- The "dip" in the obtained network strength at the optimum dose was also observed in full-scale testing of both centrifugation and gravity belt thickening. The dip indicated the optimum polymer dose based on standard dewaterability testing and solids output of both dewatering and thickening devices.

- The torque itself showed similar relationship with polymer dose as the network strength did. In order for the torque to indicate the optimum polymer dose, it must be considered at an appropriate time of applying the rotational speed. This time was found to be around 40 seconds of shear applications.

- The network strength of the residuals could be altered through changing the mixing conditions (mixing time and mixing intensity) and changing the polymer type. The network strength increased with increasing the degree of polymer structure.

- The peak of the rheograms exhibited almost a linear relationship with the polymer dose.

- Additional research is necessary test the technology in full-scale monitoring and controlling polymer dose at the optimum level and incorporating the technology in dewatering automation packages. The effect of residuals characteristics such as cations and extracellular polymeric substances that are important in residuals dewatering on network strength must also be understood.

APPENDIX A

USE OF CONCENTRIC RHEOMETERS

A-1 Materials and Methods

Rheology measurements for this part used Rotovisco RV20 with an M5 (Thermo Haake, Ltd.) sensor head, which is typical of controlled concentric rheometry. This device is capable of measuring shear stress at different controlled shear rates. This viscometer was used before in a previous WERF project conducted by the University of Delaware, and was borrowed from Dr. Dentel. The instrument was shipped from the University of Delaware to Duke University for experimentations. MVIIP sensor was used with 55 mL sludge sample. These sensors have grooved walls. All rheology measurements were obtained at a constant temperature of 25°C.

Abu-Orf and Dentel (1999) used this instrument and experienced difficulties with reproducibility of measurements. Lack of reproducibility was attributed to use of 40 mL of sludge sample when using MVIP sensor or use of 55 mL when using MVIIP sensor following conditioning of 500 mL of sludge. In this case different flocs were used in each measurements due to sub-sampling. To overcome this difficulty when using MVIIP sensor, a total volume of 55 mL of sludge was conditioned with the appropriate polymer dosage and the whole sample was used for measurements. Following polymer addition, samples were mixed by hand using a metal rod. Four replicates were used for rheology measurements where the shear rate to the sludge was increased from 0-100 in 90 seconds. These values were previously found to be appropriate for municipal sludges (Abu-Orf and Dentel, 1999; Campbell and Crecuolo, 1982). The instrument automatically measures the shear stress in (Pa) and the shear rate (1/sec), and reports the time needed to reach the recorded shear rate as well as the apparent viscosities of the samples. The ability to measure floc strength is important in sludge applications. Previous research has focused on measuring sludge floc strength indirectly through measuring other parameters such as floc size, floc density, and dewaterability of sludge. Direct measurements of floc strength as reported in the literature are not applicable to sludge, especially if conditioned. In this paper we will refer to raw or conditioned sludge suspension as "network" and thus measurements will indicate the network strength, which is most important in sludge dewaterability applications.

A-2 Results

Figure A-1 shows rheograms (shear stress vs. shear rate) for unconditioned sludge sample (a) and four different polymer dosages (b, c, d, e). In each graph four replicate measurements are reported. Although the whole sample was used for the rheology measurements, reproducibility is still an issue except for samples that were conditioned with 10g/kg (shown in graph c). As previously discussed in Chapter 1.0, Section 1.5.1, comparison of equations 1.5.1-9 and 1.5.1-10 shows that the area under a shear stress-shear rate rheogram of a non-Newtonian Bingham plastic fluid represents the rate of energy dissipation per unit volume, which indicates the total network strength of the residuals. The rheogram area will not be linearly proportional to the rate of energy dissipation as is the case for Newtonian fluids.

Figure A-2 shows the area under the rheograms (kg/m-sec3) at two different shear rates (50, and 75 sec^{-1}). Note the large error bars, which make it difficult to analyze the data. Nevertheless, if we look at the average area or network strength, we note that increasing the polymer dose lead to increases in the network strength. At 10 g/kg DS polymer dose, the area and thus the network decreased to a lower value and started to increase again as the polymer dose increased to 12 g/kg DS.

The dewaterability data obtained by the CST, filtrate volume, and the cake solids after 10 minutes filtration (Figures A-3, 4, 5) indicated that the optimum polymer dose was at 12 g/kg. Since the polymer dose was increased in increments of two, it is possible that the real optimum dose was at 11 g/kg DS and was missed during the experiments. Thus, the decrease in the network strength would have continued at 11 g/kg DS, and dewaterability data also would have indicated better dewaterability compared to 10 g/kg DS.

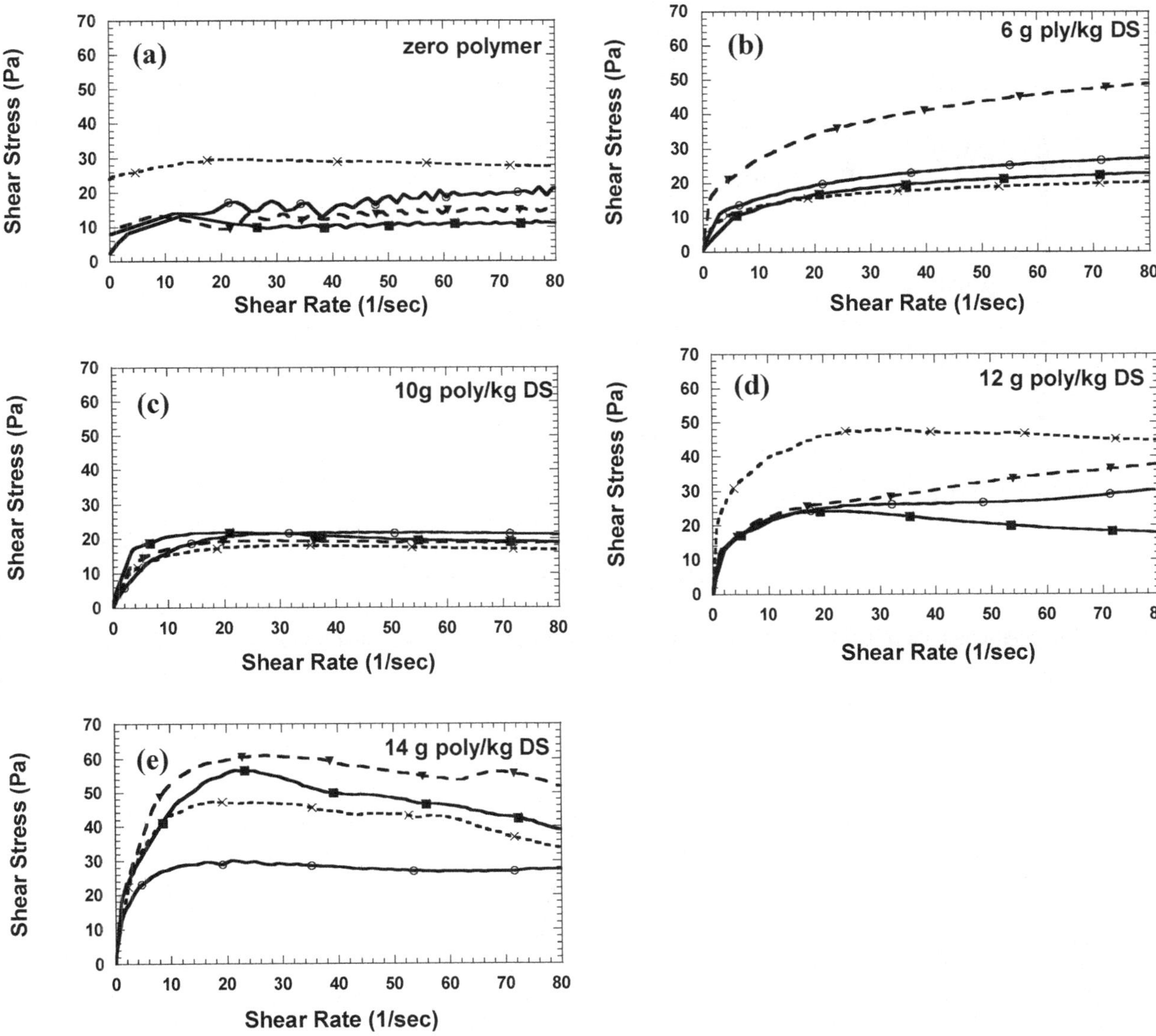

Figure A-1. Rheograms Generated Using Haake Instrument for Four Polymer Doses. Anaerobically Digested Sludge was Used.

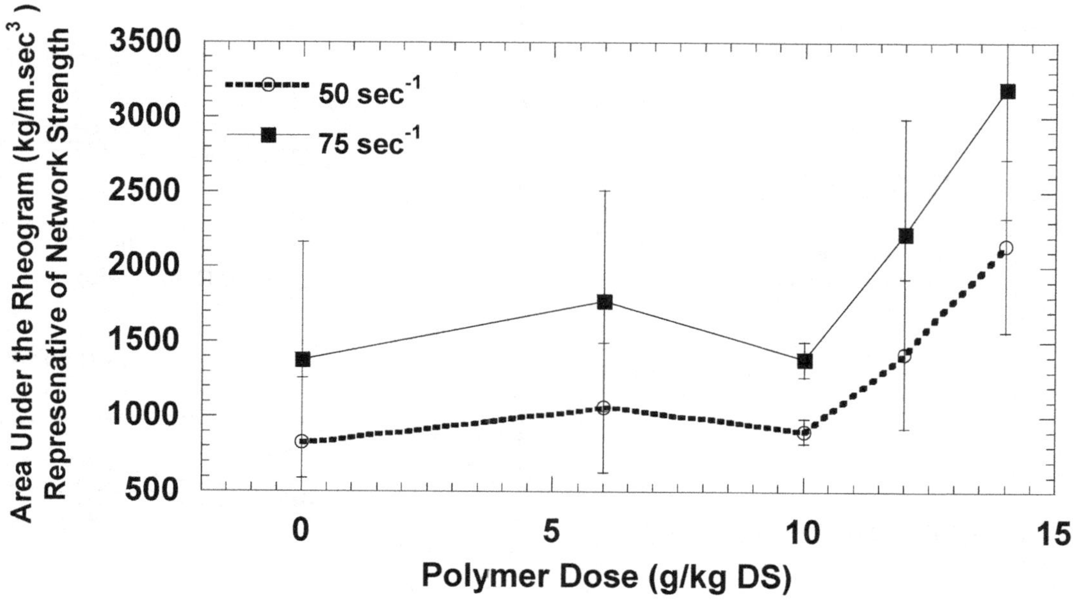

Figure A-2. Suspension's Strength Represented by Energy in Flocs at Different Shear Rates.

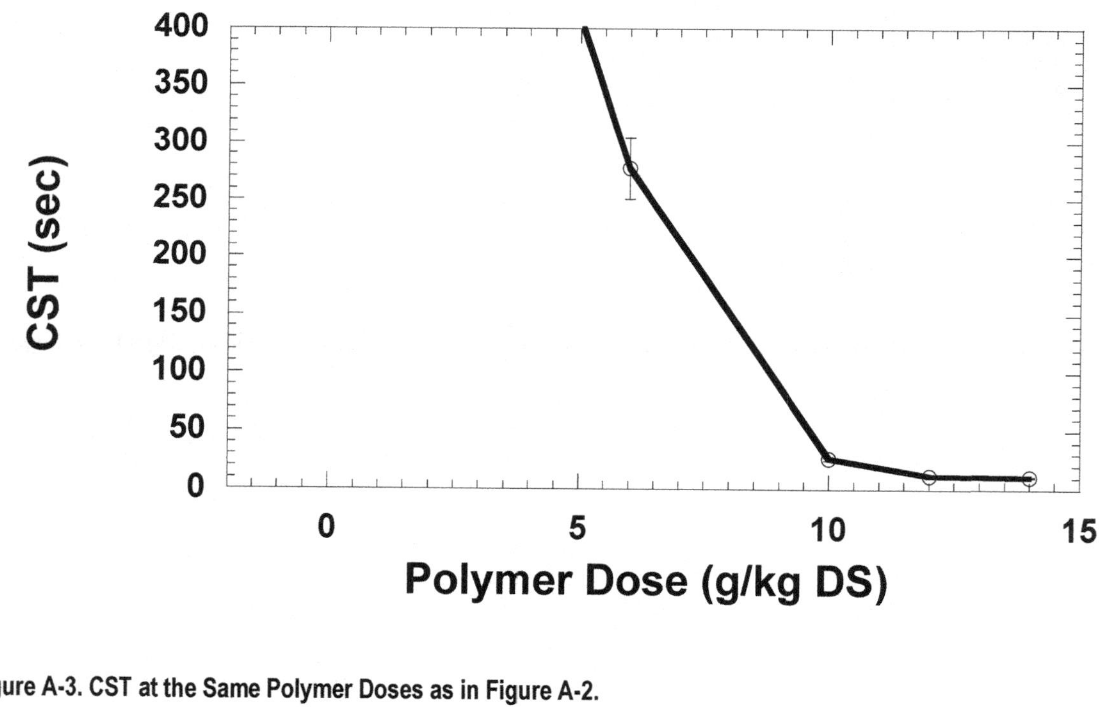

Figure A-3. CST at the Same Polymer Doses as in Figure A-2.

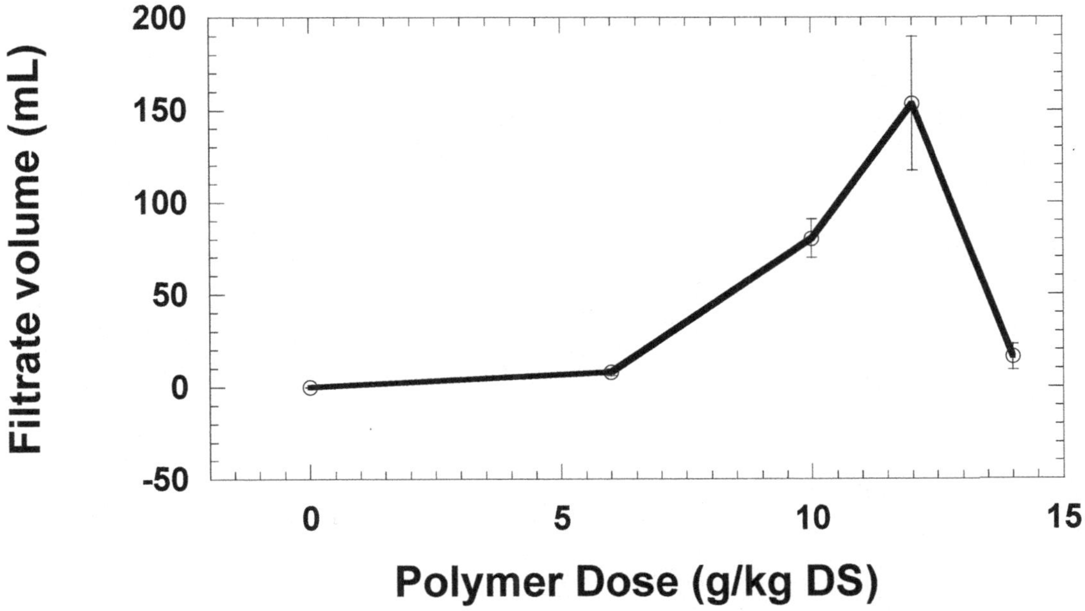

Figure A-4. Filtrate Volume After 10 Minutes of Filtration Conditioned with the Same Polymer Samples as in A-2.

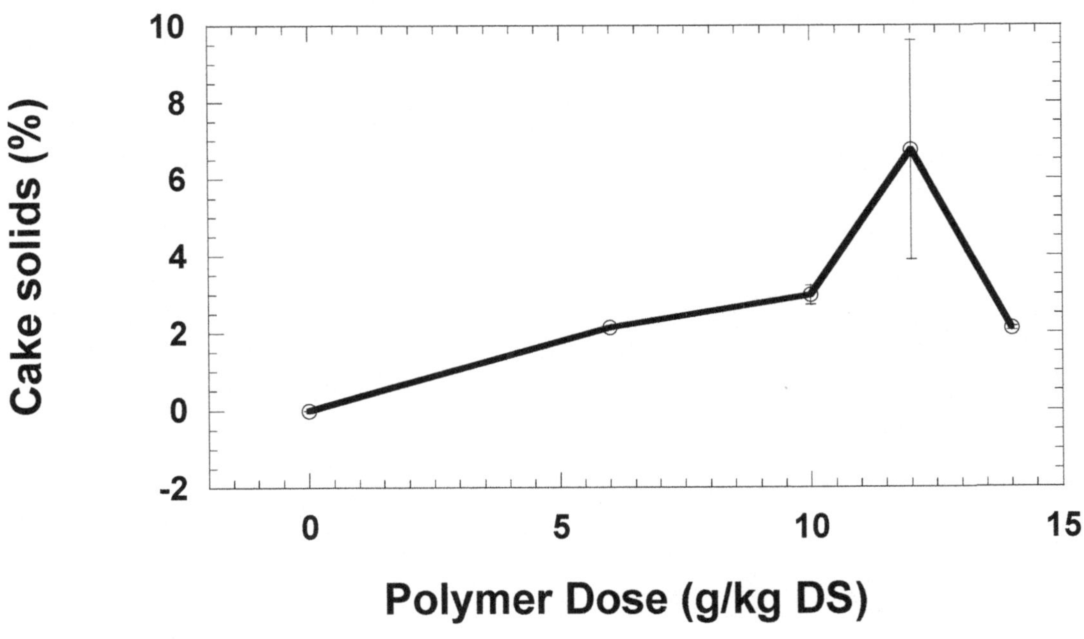

Figure A-5. Cake Solids After 10 Minutes of Filtration Conditioned with the Same Polymer Samples as in A-2.

REFERENCES

Abu-Orf, M.M.; Walker, C.A.; Dentel, S.K. Centrate viscosity for continuous monitoring of polymer feed in dewatering applications. Accepted for Publication in *Advances In Environmental Research,* 2002.

Abu-Orf, M.M.; Dentel, S.K. Rheology as a tool for polymer dose assessment and control. *J. Environ. Eng.*, ASCE. **1999,** 125(12), 1133-1141.

Abu-Orf, M.M.; Dentel, S.K. Polymer dose assessment using the streaming current detector. *Water Environ. Res.*, **1997a**, 69, No. 6, 1075-1085.

Abu-Orf, M.M., Dentel, S.K. Effect of mixing on the rheological characteristics of conditioned sludge: full-scale studies. *Water Sci. Technol.* **1997b,** 136(11), 51-60.

APHA, American Public Health Association, American Water Works Association, and the Water Environment Federation. *Standard Methods for the Examination of Wate and Wastewater.* 19th ed. Washington, D.C. 1995.

Campbell, H.W.; Crescuolo, P.J. The use of rheology for sludge characterization. *Water Sci. Technol.* **1982,** 14, 475-489.

Cheremisinoff, N.P. Chapter 34: Torque Rheometer Technology and Instrumentation, Chung, J. *Encyclopedia of Fluid Mechanics: Rheology and Non-Newtonian Flows.* Gulf Pub. Co., Book Division: Houston, 1986.

Dentel, S.K., Abu-Orf, M.M**.,** Griskowitz, N.J. *Polymer Characterization and Control in Biosolids Management.* Alexandria, VA: Water Environment Research Foundation (Publication D43007; 220 pp.), 1995.

Dentel, S.K. and Abu-Orf, M.M. Application of the Streaming Current Detector in Sludge Conditioner Selection and Control. *Water Sci. Technol.*, **1993,** 28, 169-179.

Glasgow, L.A.; Hsu, J.P. An experimental study of floc strength. *AICHE Journal.* **1982,** 28(5), 779-785.

Hannah, S.A.; Cohen, J.M.; Robeck. Measurement of floc strength by particle counting. *J. AWWA.* **1967,** Jan, 843-858.

Hiemenz, P.C.; Rajagopalan, R. *Principles of colloid and surface chemistry* – 3rd ed. Marcel Dekker Inc.: New York. 1997.

Higgins, M.J.; Novak, J.T. The effect of cations on the settling and dewatering of activated sludges: Laboratory results. *Water Environ. Res.* **1997a,** 69(2). 215-224.

Higgins, M.J.; Novak, J.T. Dewatering and settling of activated sludges: The case for using cation analysis. *Water Environ. Res.* **1997b,** 69(2). 225-232.

Kopp, J.; Dichtl, N. Influence of the free water content on the dewaterability of sewage sludges. *Wat. Sci. Technol.* **2001,** 44(10), 177-183.

Langer, S.J.; Klute, R.; Hahn, H.H. Mechanisms of floc formation in sludge conditioning with polymers. *Water Sci. Technol.* **1994,** 30(8), 129.

Michaels, S.A.; Bolger, C.J. The plastic flow behavior of flocculated kaolin suspensions. *Ind. Eng. Chem. Fundam.* **1962,** 1(3), 153-162.

Novak, J.T. The effect of mixing on the performance of sludge conditioning chemical. *Water Supply.* **1991,** 8, 53-60.

Novak, J.T.; Bandak, N. Chemical conditioning and the resistance of sludges to shear. *J. WPCF.* **1989,** 61(3), 327-332.

Novak, J.T.; Bandak, N. The effect of shear on the dewatering of water treatment residuals. *J. AWWA.* **1994,** Nov, 84-91.

Novak, J.T.; Haugan, B.E. Chemical conditioning of activated sludge. *J. Environ. Eng.* **1979,** 105(5), 993-1008.

Novak, J.T.; Love, N.G.; Smith, M.L.; Wheeler, E.R. The effect of cationic salt addition on the settling and dewatering properties of an industrial activated sludge. *Water Environ. Res.* **1998,** 70(5), 984-996.

Novak, J.T.; Prendeville, J.F.; Sherrard, J.H. Mixing intensity and polymer performance in sludge dewatering. *J. Environ. Eng.* **1988,** 114(1), 190-1197.

Otsubo, Y. Flocculation of colloids by soluble polymers and its effect on rheology. *Heterogeneous Chemical Reviews.* **1996**, 3, 327-349.

Tambo, N.; Hozumi, H. Physical characteristics of flocs – II. strength of floc. *Water Res.* **1979,** 13, 421-427.

Vesilind, P.A. The role of water in sludge dewatering. *Water Environ. Res.* **1994,** 66(1), 4-11.

Yen, P.; Chen, L.C.; Chien, C.Y.; Wu, R.; Lee, D.J. Network strength and dewaterability of flocculated sludge. *Water Res.* **2002,** 36, 539-550.

Yeung, A.K.C; Pelton, R. Micromechanics: A new approach to studying the strength and breakup of flocs. *J. Colloid Interface Sci.* **1996**, 184, 579-585.

WASTEWATER UTILITY

Alabama
Montgomery Water Works & Sanitary Sewer Board

Alaska
Anchorage Water & Wastewater Utility

Arizona
Gila Resources

Glendale, City of, Utilities Department

Mesa, City of

Peoria, City of

Phoenix Water Services Department

Pima County Wastewater Management

Arkansas
Little Rock Wastewater Utility

California
Calaveras County Water District

Central Contra Costa Sanitary District

Contra Costa Water District

Crestline Sanitation District

Delta Diablo Sanitation District

Dublin San Ramon Services District

East Bay Dischargers Authority

East Bay Municipal Utility District

El Dorado Irrigation District

Fairfield-Suisun Sewer District

Fresno Department of Public Utilities

Irvine Ranch Water District

Las Virgenes Municipal Water District

Livermore, City of

Lodi, City of

Los Angeles, City of

Los Angeles County, Sanitation Districts of

Napa Sanitation District

Orange County Sanitation District

Palo Alto, City of

Riverside, City of

Sacramento Regional County Sanitation District

San Diego Metropolitan Wastewater Department, City of

San Francisco, City & County of

San Jose, City of

Santa Barbara, City of

Santa Rosa, City of

South Bayside System Authority

South Coast Water District

South Orange County Wastewater Authority

Union Sanitary District

West Valley Sanitation District

Colorado
Boulder, City of

Colorado Springs Utilities

Littleton/Englewood Water Pollution Control Plant

Metro Wastewater Reclamation District, Denver

Connecticut
The Mattabasset District

New Haven, City of, WPCA

District of Columbia
District of Columbia Water & Sewer Authority

Florida
Broward, County of

Fort Lauderdale, City of

Gainesville Regional Utilities

JEA

Miami-Dade Water & Sewer Authority

Orange County Utilities Department

Orlando, City of

Reedy Creek Improvement District

Seminole County Environmental Services

St. Petersburg, City of

Stuart Public Utilities

Tallahassee, City of

Tampa, City of

Toho Water Authority

West Palm Beach, City of

Georgia
Atlanta Department of Watershed Management

Augusta, City of

Clayton County Water Authority

Cobb County Water System

Columbus Water Works

Fulton County

Gwinnett County Department of Public Utilities

Macon Water Authority

Savannah, City of

Hawaii
Honolulu, City and County of

Illinois
American Bottoms Wastewater Treatment Plant

Greater Peoria Sanitary District

Kankakee River Metropolitan Agency

Metropolitan Water Reclamation District of Greater Chicago

North Shore Sanitary District

Wheaton Sanitary District

Indiana
Fort Wayne, City of

Iowa
Ames, City of

Cedar Rapids Wastewater Facility

Des Moines Metro Wastewater Reclamation Authority

Iowa City, City of

Kansas
Johnson County Unified Wastewater Districts

Unified Government of Wyandotte County/Kansas City, City of

Kentucky
Louisville & Jefferson County Metropolitan Sewer District

Louisiana
Sewerage & Water Board of New Orleans

Maine
Bangor, City of

Portland Water District

Maryland
Anne Arundel County Bureau of Utility Operations

Howard County Department of Public Works

Washington Suburban Sanitary Commission

Massachusetts
Boston Water & Sewer Commission

Upper Blackstone Water Pollution Abatement District

Michigan
Ann Arbor, City of

Detroit, City of

Holland Board of Public Works

Lansing, City of

Owosso Mid-Shiawassee County WWTP

Saginaw, City of

Wayne County Department of Environment

Wyoming, City of

Minnesota
Rochester, City of

Western Lake Superior Sanitary District

Missouri
Independence, City of

Kansas City Missouri Water Services Department

Little Blue Valley Sewer District

Nebraska
Lincoln Wastewater System

Nevada
Henderson, City of

New Jersey
Passaic Valley Sewerage Commissioners

New York
New York City Department of Environmental Protection

Rockland County Solid Waste Management Authority/ Sewer District

North Carolina
Cary, Town of

Charlotte/Mecklenburg Utilities

Durham, City of

Metropolitan Sewerage District of Buncombe County

Orange Water & Sewer Authority

Ohio
Akron, City of

Butler County Department of Environmental Services

Columbus, City of

Metropolitan Sewer District of Greater Cincinnati

Northeast Ohio Regional Sewer District

Summit, County of

Oklahoma
Tulsa, City of

Oregon
Clean Water Services

Eugene/Springfield Water Pollution Control

Water Environment Services

Pennsylvania
Philadelphia, City of

University Area Joint Authority, State College

South Carolina
Charleston Commissioners of Public Works

Mount Pleasant Waterworks & Sewer Commission

Spartanburg Sanitary Sewer District

Tennessee
Cleveland, City of

Knoxville Utilities Board

Murfreesboro Water & Sewer Department

Nashville Metro Water Services

Texas
Austin, City of

Dallas Water Utilities

Denton, City of

El Paso Water Utilities

Fort Worth, City of

Gulf Coast Waste Disposal Authority

Houston, City of

San Antonio Water System

Trinity River Authority

Utah

Salt Lake City Corporation

Virginia

Alexandria Sanitation Authority

Arlington, County of

Fairfax County Virginia

Hampton Roads Sanitation District

Henrico, County of

Hopewell Regional Wastewater Treatment Facility

Loudoun County Sanitation Authority

Lynchburg Regional WWTP

Prince William County Service Authority

Richmond, City of

Rivanna Water & Sewer Authority

Washington

Edmonds, City of

Everett, City of

King County Department of Natural Resources

Seattle Public Utilities

Sunnyside, Port of

Yakima, City of

Wisconsin

Green Bay Metro Sewerage District

Madison Metropolitan Sewerage District

Milwaukee Metropolitan Sewerage District

Racine, City of

Sheboygan Regional Wastewater Treatment

Wausau Water Works

Australia

South Australian Water Corp.

Sydney Water Corp.

Water Corp. of Western Australia

Canada

Greater Vancouver Regional District

Toronto, City of, Ontario

Winnipeg, City of, Manitoba

Mexico

Servicios de Agua y Drenaje de Monterrey, I.P.D.

New Zealand

Watercare Services Limited

Singapore

Singapore Public Utilities Board

United Kingdom

Yorkshire Water Services Limited

STORMWATER UTILITY

California

Los Angeles, City of, Department of Public Works

Monterey, City of

Santa Rosa, City of

Colorado

Boulder, City of

Georgia

Griffin, City of

Iowa

Cedar Rapids Wastewater Facility

Des Moines Metro Wastewater Reclamation Authority

Kansas

Overland Park, City of

Kentucky

Louisville & Jefferson County Metropolitan Sewer District

Maine

Portland Water District

Minnesota

Western Lake Superior Sanitary District

North Carolina

Cary, Town of

Charlotte, City of, Stormwater Services

Pennsylvania

Philadelphia, City of

Tennessee

Chattanooga Storm Water Management

Washington

Seattle Public Utilities

STATE

Arkansas Department of Environmental Quality

Fresno Metropolitan Flood Control District, Calif.

Urban Drainage & Flood Control District, Colo.

CORPORATE

ADS Environmental Services

The ADVENT Group Inc.

Alan Plummer & Associates

Alden Research Laboratory

Alpine Technology Inc.

Aqua-Aerobic Systems, Inc.

Aquateam–Norwegian Water Technology Centre A/S

BaySaver Inc.

BioVir Laboratories, Inc.

Black & Veatch

Boyle Engineering Corporation

BPR CSO

Brown & Caldwell

Burns & McDonnell

CABE Associates Inc.

The Cadmus Group

Camp Dresser & McKee Inc.

Carollo Engineers Inc.

Carpenter Environmental Associates Inc.

CDS Technologies Inc.

Chemtrac Systems Inc.

CH2M HILL

Damon S. Williams Associates, LLC

David L. Sheridan, P.C.

Earth Tech Inc.

Eco-Matrix

Environmental Engineers International

EMA Inc.

The Eshelman Company Inc.

Finkbeiner, Pettis, & Strout Inc.

Freese & Nichols, Inc.

Frontier Geosciences Inc.

ftn Associates Inc.

Fuss & O'Neill, Inc.

Gannett Fleming Inc.

Golder Associates Inc.

Greeley and Hansen LLC

Hazen & Sawyer, P.C.

HDR Engineering Inc.

HNTB Corporation

HydroQual Inc.

Ingersoll-Rand Energy Systems

Insituform Technologies Inc.

Institute for Environmental Technology & Industry, Korea

Jacobson Helgoth Consultants Inc.

Jason Consultants Inc.

Jordan, Jones, & Goulding Inc.

KCI Technologies Inc.

Kelly & Weaver, P.C.

Kennedy/Jenks Consultants

Komline Sanderson Engineering Corporation

Lawler, Matusky & Skelly Engineers, LLP

Limno-Tech Inc.

Lombardo Associates Inc.

Malcolm Pirnie Inc.

McKim & Creed

MEC Analytical Systems Inc.

Metcalf & Eddy Inc.

MWH

New England Organics

Odor & Corrosion Technology Consultants Inc. (OCTC)

ONDEO Degremont Inc.

Oswald Green, LLC

PA Government Services Inc.

Parametrix Inc.

Parsons

Post, Buckley, Schuh & Jernigan

R&D Engineering/Conestoga Rover & Associates

The RETEC Group

RMC, Inc. (Raines, Melton & Carella)

R.M. Towill Corporation

Ross & Associates Ltd.

Royce Technologies

SAIC Maritime Technical Group

Sear-Brown

Short Elliott Hendrickson, Inc.

Stormwater Management, Inc.

Synagro Technologies, Inc.

Tetra Tech Inc.

Trojan Technologies Inc.

URS Corporation

Veolia Water NATC

Wade-Trim Inc.

Weston Solutions Inc.

Woodard & Curran

WRc/D&B, LLC

WWETCO, LLC

Zoeller Pump Company

INDUSTRY

American Electric Power

ChevronTexaco Energy Research & Technology Company

The Coca-Cola Company

Dow Chemical Company

DuPont Company

Eastman Kodak Company

Merck & Company Inc.

ONDEO Services

Procter & Gamble Company

PSEG Services Corp.

Severn Trent Services Inc.

Shell Global Solutions

Thames Water Plc

United Water Services LLC

Note: List as of 1/31/04

WERF Product Order Form

As a benefit of joining the Water Environment Research Foundation, subscribers are entitled to receive one complimentary copy of all final reports and other products.Additional copies are available at cost (usually $10).To order your complimentary copy of a report, please write "free" in the unit price column.WERF keeps track of all orders.If the charge differs from what is shown here, we will call to confirm the total before processing.

Name Title

Organization

Address

City State Zip Code Country

Phone Fax Email

Stock #	Product	Quantity	Unit Price	Total
			Postage & Handling	
			VA Residents Add 4.5% Sales Tax	
			Canadian Residents Add 7% GST	
			TOTAL	

Method of Payment:

(All orders must be prepaid.)

❑ Check or Money Order Enclosed

❑ Visa ❑ Mastercard ❑ American Express

Account No. Exp. Date

Signature

Shipping & Handling:

Amount of Order	United States	Canada & Mexico	All Others
Up to but not more than:	Add:	Add:	Add:
$20.00	$5.00*	$8.00	50% of amount
30.00	5.50	8.00	40% of amount
40.00	6.00	8.00	
50.00	6.50	14.00	
60.00	7.00	14.00	
80.00	8.00	14.00	
100.00	10.00	21.00	
150.00	12.50	28.00	
200.00	15.00	35.00	
More than $200.00	Add 20% of order	Add 20% of order	

*minimum amount for all orders

Note: Please make checks payable to the Water Environment Research Foundation.

To Order (Subscribers Only):

Log on to www.werf.org and click on the "Product Catalog."

Phone: (703) 684-2470
Fax: (703) 299-0742.

WERF
Attn: Subscriber Services
635 Slaters Lane
Alexandria, VA 22314-1177

To Order (Non-Subscribers):

Non-subscribers may be able to order WERF publications either through WEF (www.wef.org) or IWAP (www.iwapublishing.com).Visit WERF's website at www.werf.org for details.

IWA Publishing's authorised EU representative for General Product Safety Regulations is Diane D'Arras, 15 rue Duret, 75116 Paris, France, e-mail: safety@iwap.co.uk.

Printed and bound by CPI Group (UK) Ltd, Croydon, CR0 4YY
17/09/2026
02218798-0008